SpringerBriefs in Crystallography

SpringerBriefs in Crystallography, published under the auspices of the International Union of Crystallography, aims at presenting highly relevant, concise monographs with an intermediate scope between a topical review and a full monograph. Areas of interest include chemical crystallography, crystal engineering, crystallography of materials (ceramics, metals, organometallics, functional materials), instrumentation, mathematical crystallography, mineralogical crystallography, physical properties of crystals, structural biology and related fields.

SpringerBriefs present succinct summaries of cutting-edge research and practical applications covering a range of content from professional to academic and featuring compact volumes of 50 to 125 pages.

- A timely report of state-of-the art experimental techniques and instrumentation
- New computation algorithms or theoretical approaches
- A bridge between new research results, as published in journal articles, and a contextual literature review
- A snapshot of a hot or emerging topic
- An in-depth case study
- A presentation of core concepts that students must understand in order to make independent contributions

Briefs are characterized by fast, global electronic dissemination, standard publishing contracts, standardized manuscript preparation and formatting guidelines, and expedited production schedules.

Publications in these series help support the Outreach and Education Fund for the International Union of Crystallography.

More information about this series at https://link.springer.com/bookseries/16236

Piero Macchi

Quantum Crystallography: Expectations vs Reality

 Springer

Piero Macchi (ID)
Department of Chemistry, Materials
and Chemical Engineering
Politecnico di Milano
Milan, Italy

ISSN 2524-8596   ISSN 2524-860X   (electronic)
SpringerBriefs in Crystallography
ISBN 978-3-030-95640-0   ISBN 978-3-030-95641-7   (eBook)
https://doi.org/10.1007/978-3-030-95641-7

This Springer imprint is published by the registered company Springer Nature Switzerland AG
The registered company address is: Gewerbestrasse 11, 6330 Cham, Switzerland

Contents

Chapter 1
Introduction

Quantum crystallography deals with the application of quantum theory in crystallography.

In this SpringerBrief in crystallography, we will not treat the fundamentals of crystallography or quantum mechanics. For that purpose, the reader is invited to consult general-purpose textbooks on each subject (like Giacovazzo et al. 2016 for crystallography or Griffiths, 2016 for quantum mechanics), as well forthcoming books on quantum crystallography (Massa and Matta 2021; Macchi 2022), whereas modern applications are reported in recent reviews (Grabowsky et al. 2017, 2020; Genoni et al. 2018; Genoni and Macchi 2020). This mini review, instead, deals with a thorough comparison between the expected outcome of quantum mechanical treatment of crystallography and the state of the art of the research in this field. In other words, the purpose of this review is a "fact-checking" of the progresses in the field with respect to the promises and the potential, from the beginning of this research field until today. Most of the examples reported deal with applications on organic and organometallic crystals, but the research also concerns inorganic materials, minerals, metals and alloys, as well as macromolecular crystals.

In a recent review, the *connubium* between crystallography and quantum mechanics was illustrated (Macchi 2020). The link between the two disciplines is long standing and much tighter than scientists normally think. The entanglement started quite soon after the first experiments of X-ray diffraction carried out on crystals (Friedrich et al. 1912; Bragg and Bragg 1913). In fact, it became clear to many quantum physicists that the emerging technique was suitable to solve urgent and fundamental problems of quantum mechanics: What is the electronic structure of an atom? How atoms connect with each other to form molecules or extended solids? To answer these questions, one needed a radiation with very short wavelength, able to penetrate a material and reveal fine and small details, that any other radiation type known at that time could not visualize. The crystal is the hosting matrix able to freeze the atoms enough to make them observable with the necessary precision and, at the same time, acts as a lens that magnifies the features of a single atom or molecule by means of its intrinsic periodic homogeneity.

© The Author(s), under exclusive license to Springer Nature Switzerland AG 2022
P. Macchi, *Quantum Crystallography: Expectations vs Reality*,
SpringerBriefs in Crystallography, https://doi.org/10.1007/978-3-030-95641-7_1

On the other side, crystallography required models to exhaust the increasing flow of information possible thanks to the new X-ray diffraction technique, which disruptively modified the perspective of crystallography. Instead of studying the external morphology of a crystal, the twentieth century crystallographers became interested in its inner content and adopted a quantum mechanical perspective. However, the special "microscope" available with X-ray diffraction required a well-defined hypothesis of the shape of atoms because the scattered intensities alone do not provide a solution of the puzzle and cannot reconstruct the image like in a normal optical microscope equipped with true lenses. Quantum mechanics provided these tools.

Since then, crystallography and quantum mechanics marched together for long time, investigating several aspects of the atomic and electronic structure of matter. They still do progress together today, although this is probably not what scientists normally perceive. Many non-crystallographers are probably unaware of the historical connection with quantum mechanics and believe that crystallography is simply a technique (instead of a science). Even many crystallographers ignore how much quantum mechanics they use on a daily base when solving a crystal structure or when observing optical or electronic properties of a crystal. By showing what is the scope and purpose of quantum crystallography and by highlighting what could be exploited with this discipline, this review may teach many crystallographers what else they can do with the information and data they routinely collect and extend the scope of their research.

Last but not least, theoretical chemists may also appreciate that spectroscopy is not the only experimental branch of science able to provide valuable quantum mechanical information for theoretical methods. Currently, results of crystallographic studies are mainly used by quantum chemists only for validation of theoretical structural predictions, without any attempt to find a deeper interplay and often with poor scrutiny of the actual significance of the experimental measurements.

This review is structured as follows. First, the interplay between crystallography and quantum mechanics is described in general details, through a historical perspective. The focus is then on electron distribution, one of the reasons why the two fields embraced. From electron distribution, the chemical bonding is discussed, following the link established through a prestigious series of Gordon Research Conferences[1] and meetings like the European Charge Density Meetings.[2] Then we focus on how to calculate crystal wavefunctions or density matrices from diffraction data. This connects the research on electron charge density in position space with the research on electron charge density in momentum space, in keeping with the dual representation in quantum mechanics. This reflects another series of traditional conferences in the field, the *Sagamore conference on charge, spin and momentum density* (later

---

[1] The series of Gordon Conferences on *Electron Distribution and Chemical Bonding* started in 1978 (proposed by Philip Coppens and Vedene Hollet Smith who were also the chairs).
    See https://www.grc.org/electron-distribution-and-chemical-bonding-conference
[2] The first edition was chaired by Claude Lecomte in 1996.

named *Sagamore conference on quantum crystallography*).[3] In Chap. 6, the extension of quantum crystallography beyond the non-relativistic quantum mechanics will be discussed, within the scope of the standard model of physics. Finally, the implication of quantum crystallography for the everyday life of a crystallographer or a quantum chemist will be discussed.

Each chapter starts with an overview in simple terms, whereas more in depth discussion follows, especially in Chaps. 3–5, although the fundamentals of the theories and the machinery of the methods are presented in detail elsewhere (Macchi 2022).

---

[3] This series of conferences started in 1964, at the urging of Richard J. Weiss. For the history of this conference, see Cooper (2016).

Chapter 2
How Does Quantum Mechanics Jump in the Field of Crystallography?

One may be tempted to say that quantum theory deals with everything, because it is the paradigm of physics that interprets the nature, distribution, and behavior of fundamental particles, which constitute matter and therefore the entire universe. This is certainly correct, because an important principle correlates the quantum mechanical description of sub-nanometric particles with the classical physics at macroscopic level. It is known as the *correspondence principle* (Bohr 1920), according to which the quantum physics must approach classical physics at a large-scale. Nevertheless, quantum mechanics is undoubtedly linked to the realm of electrons and to atoms or molecules as the smallest objects constituted by electrons. Larger molecules, like proteins, polymers, nano-threads, layers and, eventually, crystals, are extended systems, for which the quantum treatment remains essential, even though they may form large scale portions of matter and feature macroscopic properties (for example the color, the hardness, the conductivity, the refraction of light, etc.).

Crystals are *periodically homogeneous* objects, meaning that there is a unit (one atom, a group of atoms/ions, one or more molecules, etc.) that periodically and identically replicates in space. The interactions among atoms or molecules in a crystal as well as the interactions with external fields are part of the scope of crystallography. However, crystallography is often intended only as a powerful microscopic technique. In fact, among the properties of a crystal, the diffraction of a radiation with sufficiently short wavelength enables reconstructing the image of the nano-sized unit cell, with a spatial resolution that can hardly be matched by any other technique. This is thanks to the diffraction/microscopy dichotomy: a diffractometer is in fact a kind of *incomplete microscope*, where the crystal itself, in combination with a mathematical tool like the Fourier transform, acts as the projection and magnifying lens. The title of Compton's Nobel Prize lecture (*X-rays as a branch of optics*) in 1927 seamlessly express this concept (Compton, 1965). Moreover, in his book *X-rays in theory and experiment*, Compton (1935) wrote:

> The x-ray diffraction experiments [...] supply us with just the same information that would be available if we could look at the atom with an x-ray microscope.

P. Macchi, *Quantum Crystallography: Expectations vs Reality*, SpringerBriefs in Crystallography, https://doi.org/10.1007/978-3-030-95641-7_2

The fundamental law for X-ray diffraction is the Bragg law, which interprets the elastic scattering of a radiation with wavelength λ by a crystal with a periodic stacking of planes equally spaced by d_{hkl}:

$$2d_{hkl}sin\vartheta = n\lambda \tag{2.1}$$

where ϑ is the angle between the incoming radiation and the plane, equal to the angle between the diffracted radiation and the plane; and hkl is the set of Miller indices identifying the planes (or better the vector normal to the plane). The synthesis by Bragg follows from the classical Thomson scattering of the electromagnetic radiation by a charged particle. From Eq. (2.1), it is obvious that a short wavelength is necessary to observe small interplanar distances. Thus, X-rays represent the ideal radiation to observe small objects like atoms and molecules and map the distribution of electrons around them.

This unprecedented opportunity to scrutinize the extremely small-scale world of fundamental particles, and especially of electrons, did not escape the attention of early quantum physicists. It is probably not very well known that W. L. Bragg gave the inaugural talk at the Solvay conference on "electrons and phonons", held in Brussels, in 1927 (see Fig. 2.1 for the group photo). This meeting is considered the official establishment of quantum mechanics, combining the viewpoints of different schools (like Bohr's and Schrödinger's).[1] Bragg presented state-of-the-art results on the interpretation of X-ray diffracted intensities. While the most relevant outcome that far obtained concerned the solution of crystal structures of some fundamental solids (like diamond, rock salts, etc.), the appeal of Bragg's work for the quantum mechanics community consisted in the envisaged possibility to reconstruct the electronic shell distributions of atoms in crystals. Thus, crystals were interesting for their magnifying power in keeping with the above-mentioned analogy with microscopy. The crystal structures themselves were not so interesting, instead. This is not different from the modern perception by chemists of X-ray diffraction (often, but improperly, called X-ray crystallography), used to determine the structure of a *molecule*, while the *crystal* structure is often ignored.

The expectations of quantum physicists from the work of Bragg came from two seminal predictions by Debye and Compton, more than ten years earlier:

It seems to me that the experimental study of the scattered radiation, in particular from light atoms, should get more attention, since along this way it should be possible to determine the arrangement of the electrons in the atoms.

P. Debye (1915).

It is hoped that it will be possible in this manner [through X-ray diffraction] to obtain more definite information concerning the distribution of the electrons in the atoms.

A. H. Compton (1915).

---

[1] Popper (1982), for example, considered the Solvay conference as the meeting where the schism in quantum mechanics occurred.

Fig. 2.1 Group photo taken at the Solvay Conference 1927 on Electrons and Photons. The names of the participants are in the bottom line, ordered as they appear in the picture. Courtesy of the Solvay Institutes, Brussels

To better appreciate how quantum mechanics is involved, it is useful recalling its principles, starting from the postulates, i.e. the columns holding the whole quantum theory temple (see Fig. 2.2):

- The state of a system is described by a wavefunction ψ, which depends on the coordinates of electrons and nuclei in position space.
- The wavefunction is an amplitude probability for the particle, and the square of its modulus, $|\psi|^2$, is proportional to the probability of finding the particle at a position of the space.
- The wavefunction enables predicting any observable property of the system. A property is associated with a Hermitian operator[2] which acts on the wavefunction. The measurement is equivalent to sampling the operator over the particle probability distribution in a particular state.
- The position and the momentum of an electron are themselves operators, but in quantum mechanics they do not commute, meaning that the order with which the

---

[2] A Hermitian operator is represented by a symmetric matrix with real eigenvalues.

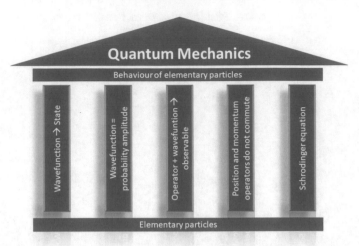

Fig. 2.2 A schematic representation of the pillars of quantum theory

observation is made is not irrelevant. For this reason, we may represent a wave-function, and any scalar or vectorial properties derived from it, in two alternative spaces: the *position space* and the *momentum space*.

- A fundamental hypothesis of quantum mechanics is the *dual* behavior of electrons as particles and waves. For this reason, the equation of motion of the electrons is the propagation of a wave (Schrödinger equation):

$$i\hbar \frac{\partial}{\partial t}\psi(\mathbf{r}, t) = \left[\frac{-\hbar^2}{2m}\nabla^2 + \widehat{V}(\mathbf{r}, t) \right]\psi(\mathbf{r}, t) = \widehat{\mathcal{H}}\psi(\mathbf{r}, t) \qquad (2.2)$$

where the wavefunction ψ depends on space coordinates and time, $\hbar$ is the Plank's constant divided by 2π and the terms in square parenthesis are the *kinetic* and *potential* energy operators, grouped in the Hamiltonian $\widehat{\mathcal{H}}$. The time independent version is:

$$\left[\frac{-\hbar^2}{2m}\nabla^2 + \widehat{V}(\mathbf{r}) \right]\psi(\mathbf{r}) = \widehat{\mathcal{H}}\psi(\mathbf{r}) = \mathcal{E}\psi(\mathbf{r}) \qquad (2.3)$$

where $\mathcal{E}$ is the energy of the system. The particle/wave dualism can be summarized as it follows: the electron is a wave which propagates according to Schrödinger equation, but when we measure it, the wavefunction collapses, and we obtain a particle. Because of postulate nr. 2, at each position we can only measure a probability to find the particle.

For crystals, the consequences of these postulates can be translated as it follows:

- The wavefunction ψ describes the state of the crystal, consisting of many electrons and nuclei.

- The wavefunction represents the probability amplitude of the electron distribution which must comply with the crystal symmetry (translational and roto/reflection). One way to construct a suitable crystal wavefunction is using Bloch's theorem (Bloch 1928). The solution of the Schrödinger equation for a periodic system is a plane wave $\psi(\mathbf{r}) = e^{i\mathbf{k}\mathbf{r}}\phi(\mathbf{r})$, where $\phi(\mathbf{r})$ is a periodic function that simply replicates the wavefunction of the unit cell along the three directions. The function $\phi(\mathbf{r})$ can be a plane wave itself or a combination of plane waves, which are inherently periodic. Alternatively, the function may be like the solutions of the Schrödinger equation for molecules, with a wavefunction which is a product of molecular orbitals, obtained as a linear combination of atomic orbitals. In periodic crystals, these orbitals are the *crystal orbitals*, in analogy with the molecular case.
- All properties of the crystals can be computed from the wavefunction, again in keeping with the crystal symmetry. The Hamilton operator is a periodic potential and kinetic energy operator, and it returns the energy of the crystal. The *scattering operator* describes the scattering of electromagnetic radiation and its expectation value of the scattering operator returns a quantity which is proportional to the measured intensity.
- Because we have two different representations, in position or in momentum space, the wavefunction and the corresponding electron density are also described in two possible spaces. This implies different measurements for the determination of the electron distribution: Bragg scattering for the position space; Compton scattering for the momentum space.
- The Schrödinger equation holds also when the potential is periodic. The Hamiltonian must therefore consider the interactions of the entire periodic system.

There is no doubt, that among these declinations of quantum mechanics postulates into the crystallography realm, the scattering operator is one of the most interesting together with the definition of a periodic probability of finding electrons.

In this respect, the link between diffraction of X-rays and quantum mechanics is straightforward. The elastic scattering of a photon is a process that involves the system in an electronic state, which remains unperturbed by the event. The emitted wave has the same energy (hence the same wavelength) of the incoming radiation and it conveys an information on the distribution of the particle in the position space. We assume a perfect coherence among waves: the inference occurs among all waves scattered along a given direction by the electrons at different positions. The operator is therefore simply a summation of wave amplitudes, modulated by the *phase difference*, due to their non-coincident origin and depending on the direction and amount of momentum transfer $\mathbf{k}$ (also called *scattering vector*) between incident and diffracted waves ($\mathbf{k} = \mathbf{k}_{\text{diffr.}} - \mathbf{k}_{\text{inc.}}$; with $|\mathbf{k}_{\text{diffr.}}| = |\mathbf{k}_{\text{inc.}}| = 2\pi/\lambda$ and $|\mathbf{k}| = 4\pi \sin\vartheta/\lambda$). The diffracted wave sums up the waves scattered by each electron, thus the operator for the scattering of one atom (consisting of n electrons) is:

$$\widehat{f} = \sum_{j=1}^{n} e^{i\mathbf{k}\cdot\mathbf{r}_j} \tag{2.4}$$

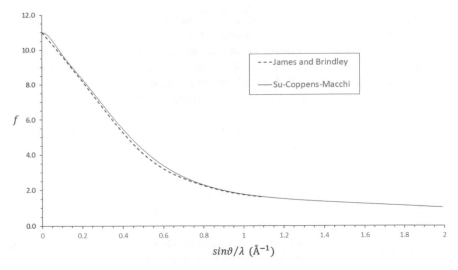

Fig. 2.3 The atomic form factor of Na. The quantity is normalized to the scattering lenght of one electron. Two calculations are shown: one from James and Brindley (1931), which is one of the first attempts to calculate with quantum mechanical methods the atomic form factor; the other applies the method from Su and Coppens (1998), Macchi and Coppens (2001). Note the small difference between the two curves, although the latter includes relativistic treatment and configuration interaction, in other words one of the most accurate theoretical level against one of the most primitive

The intensity that one can measure is proportional to the square of the expectation value of the scattering operator $\langle \widehat{f} \rangle^2$, which implies the loss of the overall phase:

$$f(\mathbf{k}) = \hat{f}$$

$$= \int \psi_{atom}^*(\mathbf{r}_1, \mathbf{r}_2, \ldots, \mathbf{r}_n) \left(\sum_{j=1}^{n} e^{i\mathbf{k}\cdot\mathbf{r}_j} \right) \psi_{atom}(\mathbf{r}_1, \mathbf{r}_2, \ldots, \mathbf{r}_n) d\mathbf{r}_1 d\mathbf{r}_2 \ldots d\mathbf{r}_n$$

$$= \int e^{i\mathbf{k}\cdot\mathbf{r}} \rho_{atom}(\mathbf{r}) dV \qquad (2.5)$$

where $\rho_{atom}(\mathbf{r})$ is the atomic electron density distribution, which is a function of the position coordinate $\mathbf{r}$ and no longer depends on the individual electron coordinates $\mathbf{r}_j$. The function f, which depends on the momentum transfer $\mathbf{k}$, is called the *atomic form factor* (see Fig. 2.3). It is a Fourier transform of the atomic electron density.

The scattering by the whole crystal can be considered as the collection of waves scattered by each atom j located at position $\mathbf{r}_j$ inside the unit cell, that produce an overall crystal wave (the *structure factor* F), modulated by the direction and the magnitude of $\mathbf{k}$.

$$F(\mathbf{k}) = \int \rho_{unit\ cell}(\mathbf{r})e^{i\mathbf{k}\cdot\mathbf{r}}dV \sim \sum_{j=1}^{N} f_j(\mathbf{k})e^{i\mathbf{k}\cdot\mathbf{r}_j} \qquad (2.6)$$

Because of the periodic nature of the crystal, the elastic scattering does not produce a continuous F. In fact, because of the interference, $F(\mathbf{k})$ is null, unless for those $\mathbf{k}$ vectors that satisfy the diffraction condition. By projecting $\mathbf{k}$ along the three crystallographic directions ($\mathbf{a}, \mathbf{b}, \mathbf{c}$), one obtains the so-called Laue conditions for obtaining diffraction (Friedrich et al. 1912):

$$\begin{aligned} \mathbf{k} \cdot \mathbf{a} &= 2\pi h \\ \mathbf{k} \cdot \mathbf{b} &= 2\pi k \\ \mathbf{k} \cdot \mathbf{c} &= 2\pi l \end{aligned} \qquad (2.7)$$

where h, k, l are three integers.

The atomistic assumption in the second part pf Eq. (2.6) enabled the solutions of crystal structures in the early days of X-ray diffraction, although the atomic form factors adopted at that time were not very accurate, at least until quantum mechanical models of atoms became available. Those calculations were feasible thanks to the method introduced by Hartree (1928a, b) to calculate atomic wavefunctions. Nowadays, those calculations are not considered as particularly accurate, especially in terms of energy of the atomic ground state. Nevertheless, the form factors calculated with the method of Hartree were seamlessly efficient for the purpose and, if compared with more modern and accurate calculations, they do not differ substantially (see Fig. 2.3).

How is it possible that a quantum mechanical model that is not accurate enough to calculate the energy of a system is anyway efficient for estimation of the atomic form factors? The errors in the energy that bother theoretical chemists mainly concern energy differences (e.g. between products and reagents in a chemical reaction, or between bound and unbound states), whereas the diffracted intensities indirectly depend on the overall energy of all electrons in an atom. Moreover, the scattering mainly depends on inner (core) electrons, more tightly bound and localized near the nucleus, that scatter up to higher diffraction angles than valence electrons. This explains why a reasonable, albeit not fully accurate, atomic quantum model is sufficient for the solution and refinement of crystal structures from X-ray diffraction experiments. As we will see in the next chapter, the structural model typically adopted is a simple sum of atoms, which does not consider the interaction among them giving rise to chemical bonds and therefore molecules/polymers. Obviously, this is a very poor model of the crystal realm, nevertheless sufficient to publish ca. 100'000 crystal

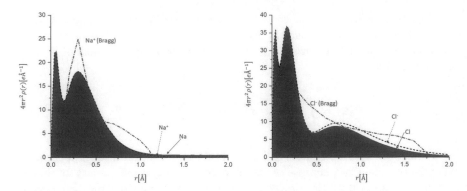

Fig. 2.4 The radial density distribution $(4\pi r^2 \rho(r))$ of Na (left) and Cl (right), determined experimentally by Bragg et al. (1922) from X-ray diffraction on NaCl crystals. For comparison, the distributions computed with accurate theoretical wavefunctions for the neutral and ionic forms of these atoms are plotted

structure per year. A scientist who solves a crystal structure is perhaps not an expert of quantum mechanics (and perhaps not even a crystallographer) and uses calculated structure factors that, although improved over the years, do not differ substantially from those adopted in the early days of the so-called X-ray crystallography. This is not enough for quantum crystallography, whose purpose is going beyond quasi-classical treatments and refine models that adhere more precisely to the realm of an electron distribution in a crystal.

Conversely, one may ask which information crystallography could return in the absence of a theoretical model. In other words, would crystallography, through the diffraction experiment, reveal the atomic electron density if we had no hypothesis on the atomic wavefunction? We may use the Bohr atom model, which assumes that electrons are grouped in shells around the nucleus, without knowing the exact wavefunction. This was in fact the state of the art in the period between 1913, when Bohr proposed his model and X-ray diffraction from crystals were first measured, and 1925, when Schrödinger proposed the wave equation, or more fairly between 1913 and 1930, when first methods to solve the Schrödinger equation for multi-electron atoms were proposed (Hartree 1928; Fock 1930). In a seminal paper, Bragg et al. (1922) attempted such a solution. For crystals like NaCl, they were able to retrieve an experimental determination of the electronic shell of Na^+ and Cl^- atoms (see Fig. 2.4). This required partitioning the measured X-ray diffracted intensities into contributions from Na and from Cl atoms, based solely on the hypothesis that the two atoms are in ionic states.

It is quite interesting to remind that the crystal structure itself of rock salts was not universally accepted in the scientific community and even many years after the first experiments, some scientists still casted doubts on the structure instead of admitting the non-existence of a separate entity like a molecule of sodium chloride. In fact, Bragg (1927) wrote that:

In sodium chloride there appear to be no molecules represented by NaCl. The equality in number of sodium and chlorine atoms is arrived at by a chess-board pattern of these atoms; it is a result of geometry and not of a pairing-off of the atoms.

However, Armstrong (1927) replied:

This statement is more than "repugnant to common sense". It is absurd to the n.....th degree, not chemical cricket. Chemistry is neither chess nor geometry, whatever X-ray physics may be. Such unjustified aspersion of the molecular character of our most necessary condiment must not be allowed any longer to pass unchallenged. A little study of the Apostle Paul may be recommended to Prof. Bragg, as a necessary preliminary even to X-ray work, especially as the doctrine has been insistently advocated at the recent Flat Races at Leeds, that science is the pursuit of truth. It were time that chemists took charge of chemistry once more and protected neophytes against the worship of false gods: at least taught them to ask for something more than chess-board evidence.

Bragg et al. (1922) anyway were able to refine a model of the radial distribution density for the two atoms in the crystal (Fig. 2.4). Albeit very approximate and affected by fundamental errors, like the number of electronic shells, their result was surprisingly close to the correct one (as we can calculate today using theoretical accurate wavefunctions for the two ions). They tentatively calculated the number of electrons per each shell, which was incorrect, but exceptional at that time. Despite this optimism, a crystallographic experiment in the absence of a solid quantum mechanical hypothesis for the electron density distribution would be insufficient to exhaust the expectations of quantum crystallography.

A final remark is necessary. In Eq. (2.6), as well as in the rest of this chapter, the dynamics has not been considered. However, atoms in crystals are not steady, but they move, albeit much less than in other aggregation states of matter. Thus, the nuclei are instantaneously displaced from their equilibrium positions, due to lattice phonons. Even at the ideal temperature of 0 K, there would be a zero-point vibration. Thus, the observed structure factors are the Fourier transform of the thermally smeared electron density $\overline{\rho}(\mathbf{r})$, i.e. averaged over all possible vibrational eigenmodes of the lattice. Assuming that each atomic electron density rigidly follows the nuclear displacement, the structure factor is more properly defined as:

$$F(\mathbf{k}) = \int \overline{\rho}_{unitcell}(\mathbf{r})e^{i\mathbf{k}\cdot\mathbf{r}}dV \sim \sum_{j=1}^{N} f_j(\mathbf{k})T_j(\mathbf{k})e^{i\mathbf{k}\cdot\mathbf{r}_j} \qquad (2.8)$$

$T_j(\mathbf{k})$ is the transform of the nuclear probability density (Debye–Waller factor, Debye 1913; Waller 1923). Under the approximation that the motion is harmonic (although not identical in all directions):

$$T_j(\mathbf{k}) = e^{-\frac{1}{2}\mathbf{k}^T\mathbf{U}_j\mathbf{k}} \qquad (2.9)$$

where $\mathbf{U}_j$ is a second order tensor, representing the squared displacement of atom j from position $\mathbf{r}_j$.

The main conclusion of this chapter is that a quantum theory is necessary to extract correct information from the experiments. Nonetheless, the diffraction experiments are, together with electron spectroscopy, the most suitable ones to demonstrate the validity of quantum theory. A failure in explaining the X-ray diffraction observations would have brought to a profound reconsideration of the theory.

In the next chapter, we will see how to improve the structural models, in keeping with a more physically grounded theory, and extract more substantial quantum mechanical information from the X-ray diffraction experiments.

Chapter 3
Beyond the Semi-classical View of Atoms and Molecules in Crystals

3.1 "Can We See the Electrons?"

This intriguing question coincides with the title of a paper published by Philip Coppens in the Journal of Chemical Education almost 40 years ago (Coppens 1984). Although it may be surprising, this question is still asked by many scientists, not only by students! As Coppens wrote, the quantum mechanical paradigm teaches us that we cannot determine the exact position of an electron at a specific time. Better reformulated, this is the Heisenberg (1927) *uncertainty principle*:

$$\Delta p \Delta q \geq \frac{1}{2}\hbar \tag{3.1}$$

where p and q are the momentum and the position coordinates of an electron, Δ is the uncertainty with which each quantity can be determined.

Apart from the philosophical interpretation, the main consequence is that we cannot expect to see the electrons and trace their trajectories in the same way as we observe and describe the motion of a tennis ball or that of a planet. The Heisenberg's principle derives from the fourth postulate of quantum mechanics (see Chap. 2), and it seamlessly affects the attempts to make such an observation like the position and simultaneously the momentum of an electron. This does not mean that one cannot carry out the two measurements on a system, but the two observations will not be associated to the same realm, because the returned value depends on the measurement sequence. This is quite different from any measurement within the limits of validity of the classical physics.

With these caveats in mind, in the experiments based on scattering we find information on the position or the momentum of electrons. What we can determine in this way, is the electron density. More precisely, the observable quantity is the *one-electron density*, namely the probability to find *any* one electron at a particular position or with a particular momentum. Thus, we must resort on a probabilistic response,

© The Author(s), under exclusive license to Springer Nature Switzerland AG 2022
P. Macchi, *Quantum Crystallography: Expectations vs Reality*,
SpringerBriefs in Crystallography, https://doi.org/10.1007/978-3-030-95641-7_3

in keeping with the second postulate of quantum mechanics. Moreover, we cannot recognize the individual electrons with scattering experiments.

Anyway, to answer in short: yes, *we can see the electrons*. Though, how?

We may rephrase the title of this Chapter: *how to map the electron density distribution around atoms in crystals*. As mentioned, this opportunity was prophetically envisaged by Debye and Compton in the early days of the X-ray diffraction technique, but it was not immediately possible. We already discussed about the attempt made by Bragg et al. (1922) to reconstruct the atomic electron shells of Na and Cl atoms from X-ray diffraction of NaCl crystals (see Fig. 2.4). This approach, unique at that time but poorly remembered in the following years, was extraordinarily innovative. The failure to exactly reconstruct the electronic shells of the atoms was due to the inherent difficulty of mapping very precisely the electron distribution in the vicinity of the nuclei, inside the core region, for which one needs a diffraction resolution which was (and still is) hard to obtain and goes well beyond the standard resolution of X-ray diffraction experiments.

Interestingly, Bragg's approach dealt also with a fundamental problem in quantum chemistry: to which atomic nucleus belong the electrons of a molecule/crystal? One cannot *label* the electron density of an atom when it is aggregated with others. It is not possible to inventory the electrons even if atoms were not covalently bonded (thus, sharing their valence) but linked through closed-shell interactions (like ionic bonds).

We can analyze the problem from two different points of views, relevant for the models of the experimental observation or first principles calculations:

(a)   *Experimentally*, we assume that the diffraction from a crystal comes from the combination of waves scattered individually by atoms; however, once the interference has occurred, the genealogy is lost, and we can no longer recognize the individual atomic waves.

(b)   *Theoretically*, we approximate a system wavefunction with the product of linear combinations of atomic (orbital) functions and from these we calculate the electron density; however, an infinite number of wavefunctions correspond to the same electron density (Gilbert 1975), thus atomic orbitals may combine in different ways to produce the observable electron density (Palke 1986).

How can we reconstruct, then, the contribution of individual atoms?

For the electronic shells of atoms in NaCl, Bragg et al. (1922) adopted an arbitrary breakdown of the diffracted intensities into atomic scattering. For a more complicated system than a binary salt, though, the problem may be exceedingly difficult. Exhaustive treatment of the methods of partitioning the electron density into atomic domains will be the subject of Chap. 4. Here, we tackle the problem with the purpose to construct a model suitable for the interpretation of X-ray diffraction experiments. The model should enable to:

(1)   assign each atom a (fractional) number of electrons and determine its charge.
(2)   identify electrons mostly contributing to the chemical bonding, the behavior of which may substantially deviate from that in unperturbed isolated atoms.

3.1.1 Ionization State of Atoms

The former goal, determining the atomic charge state, was bypassed by Bragg with an a priori assumption. In fact, determining the atomic charge from diffracted X-rays is not an easy task. In his seminal book on the optical theory of X-ray diffraction, James (1958) was particularly skeptic about the possibility to determine the atomic charges from X-ray diffraction. The atomic form factors of a neutral or an ionic electronic configuration differ only slightly (see Fig. 3.1 for the comparison between Fe and Fe^{2+}).

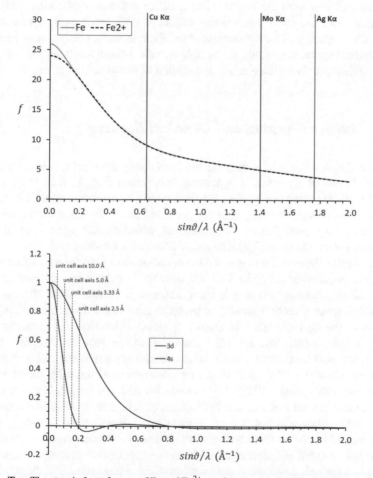

Fig. 3.1 **Top**: The atomic form factors of Fe and Fe^{2+} as a function of $sin\vartheta/\lambda$. A difference between the two curves is visible only in the region $0-0.2$ Å^{-1}. The best resolution available with some typical radiation wavelengths (Cu Kα, Mo Kα, Ag Kα) is indicated. **Bottom**: The contribution of 4s and 3d electrons to the scattering of Fe as a function of $sin\vartheta/\lambda$. The position of the first reflection measurable with different unit cell axes is marked, showing that for very small unit cells, no information about 4s orbitals is available and that it is anyway limited even for larger unit cells

The outermost electrons, on which the atomic charge depends, scatter only at extremely low diffraction angles, because of the reciprocity between direct position space and diffraction (reciprocal) space. If the crystal unit cell were not large enough, Bragg diffraction conditions would not occur at such small angles, hence the possibility to spot the presence, or absence, of outermost electrons would vanish. Moreover, even if the unit cell would be large enough to produce a few reciprocal lattice nodes at small $|\mathbf{k}|$ (thus, small $sin\vartheta/\lambda$), the intensities may be strongly affected by secondary extinction and/or absorption, hampering once again the possibility to determine with accuracy the exact electronic configuration of an atom in a crystal. In the bottom plot of Fig. 3.1, the position of the first diffraction peak is highlighted for some unit cell axes dimensions. For a lattice parameter of 10.0Å, only three reflections along the corresponding reciprocal space direction may capture, at least in part, the scattering of the 4s electron of Fe. Even including the other directions of the reciprocal space, one would not be able to collect enough reflections to estimate with sufficient precision the electron population of the 4s orbital of Fe.

3.1.2 Valence Electrons and Chemical Bonding

The second task addressed above, namely identifying the electrons involved in the chemical bonds of an atom, is apparently less demanding. In fact, if an atom is bonded to another covalently, then the distribution of electrons around its nucleus deviates from spherical, being strongly polarized along the direction of the bond. This makes the atomic form factor aspherical, affecting diffracted intensities in a non-homogeneous fashion. The difference between a bonding and a non-bonding direction in the direct space is mapped in the reciprocal space for a broad resolution range and the affected reflections are not only those at very low diffraction angle. In Fig. 3.2, we see how the atomic form factor of an oxygen atom changes along a reciprocal space direction parallel or perpendicular to the direct space O=O bond direction of the O_2 molecule. The scattering along a direction perpendicular to the bond is not distinguishable from the scattering of the atom in isolation. On the contrary, the direction parallel to the bond experiences a non-negligible distortion. This is mainly due to the interference between waves emitted by the two atoms (see for example Guinier 1963). It is evident for a single molecule as well as for a hypothetical crystal made of parallel molecules. The ample oscillations along the O-O direction, shown in Fig. 3.2, are anyway not a direct clue of bonding, as they simply depend on the distance between two sources of the scattered radiation.

However, a small but significant difference is noticeable between a model that accounts for the polarization and deformation of the atomic densities (the continuous lines) and a model that does not consider these effects (dashed lines). Again, this is true only along the O-O direction, because perpendicular to it the two models are seamlessly indistinguishable. In the top plot of Fig. 3.2, one can compare the resolution limits available with some commonly adopted laboratory sources (like

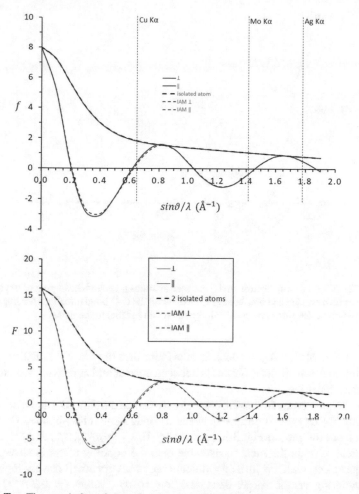

Fig. 3.2 Top: The atomic form factor of an O atom in O_2 molecule. The red curve is the form factor along a reciprocal space direction perpendicular to the O=O bond and it almost coincides with the form factor for an isolated O atom. A very accurate model (like the multipolar model) and the IAM coincide. The blue curve is the form factor along a reciprocal space direction parallel to the bond direction. The interference among the bonded atoms is evident and this makes the scattering of the atom in the molecule quite different from the scattering of the atom in isolation. Moreover, the exact curve differs more significantly from the IAM approximation. The resolution available with some typical radiation wavelengths (Cu Kα, Mo Kα, Ag Kα) is also indicated. **Bottom**: The structure factor of an O_2 molecule within a hypothetical crystal lattice with a cubic metrics but without any symmetry (hence in space group $P1$). The structure factor is extrapolated from the discrete distribution of values at diffraction angles fixed by the lattice parameters. The color codes are the same as for the top figure. One can easily appreciated that these curves have double values compared to the corresponding ones in the top graph

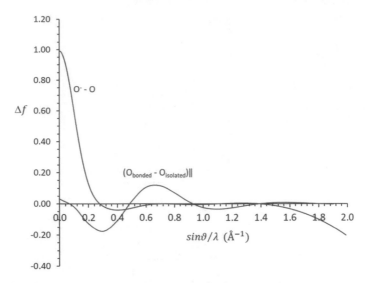

Fig. 3.3 The difference (in electron units) between anionic and neutral scattering of oxygen (blue line) and between bonded and non-bonded O atom along the O-O bond direction (red line). For the bonded atom, the scattering is evaluated along a direction parallel to the bond

tubes with Cu, Mo or Ag anodes). It is evident that the Cu Kα radiation may be insufficient to catch all the differences between correct and approximated models of the electron densities.

In Fig. 3.3, we compare the differences between an anionic (O⁻) and a neutral (O) form factor for oxygen and those between a bonded and a non-bonded O atom, as obtained from the plot in Fig. 3.2. In the range $0.2 < sin\vartheta/\lambda < 1.0 \text{Å}^{-1}$, which is for practical reasons the most measurable one, the bonding effect is stronger than the charge effect, which is instead dominant only for very small or very high values of the scattering vector. As we discussed, low $sin\vartheta/\lambda$ values ($< 0.2 \text{Å}^{-1}$) are not measurable in real circumstances because almost no Bragg reflection occurs there, whereas the high $sin\vartheta/\lambda$ values ($>1.8 \text{Å}^{-1}$) are unreachable unless using high-energy radiation, beyond the typical laboratory sources. From Fig. 3.3 it emerges quite clearly that it is easier to capture polarization effects of chemical bonding than to determine the charge of an atom.

3.1.3   The Independent Atom Model and Its Limitations

Because of the significant effect produced by the chemical bonding, a quantity is often analyzed, which is relatively simple to calculate, namely the *deformation electron density* (discussed in more details in Sect. 3.2). This function emphasizes the difference between the electron density of a molecule/crystal and that calculated by assembling spherical, neutral, and non-interacting atoms, as used in Fig. 3.2. The reference model is the so-called *promolecule*, a name introduced by Hirshfeld and Rzotkiewicz (1974). In the position space, the promolecule electron density is defined as:

$$\rho_{promolecule}(\mathbf{r}) = \sum_{i \in molecule} \rho_{spher,i}(\mathbf{r} - \mathbf{R}_i) \qquad (3.2)$$

where $\mathbf{R}_i$ is the position-vector of atom i and $\rho_{spher,i}$ is its atomic ground state electron density, as obtained from a quantum mechanical calculation. $\rho_{spher,i}$ is spherically averaged among the electronic microstates,[1] otherwise a bias would be introduced in the reference density. The promolecule model is not a novelty, though. Indeed, it coincides with the model proposed by Debye (1930) to obtain a reasonable solution of the crystal structures:

$$\rho_{unitcell}(\mathbf{r}) = \sum_{i \in unit\ cell} \rho_{spher,i}(\mathbf{r} - \mathbf{R}_i) \qquad (3.3)$$

This model is known as the *Independent Atom Model* (IAM) or the *Spherical Atom Model*. Noteworthy, the "independency" does not concern only the electron density, but also the vibrational modes of atoms, that can be treated as a collection of uncorrelated oscillators, independent from one other. As we have discussed in Chap. 2, the atomic electron density is supposed to follow the nucleus along the displacements from the equilibrium position. This assumption is very important for the construction of a proper atomic form factor, which represents the electron distribution of an atom smeared by its oscillation.

If the IAM is used not only to find the set of atomic positions that better explains the measured intensities, but also as a reference for the exact electron density, we need a way to convert the structure factors into electron densities. For this purpose, one needs to operate a Fourier inversion, i.e. transforming the wave amplitudes into the scattering matter (the electron density), thus:

$$\rho(\mathbf{r}) = \frac{1}{V_{unit\ cell}} \sum_{\mathbf{k}} F_{\mathbf{k}} e^{-i\mathbf{k}\mathbf{r}} \qquad (3.4)$$

where the $\mathbf{k}$ vectors included in the summation are those that satisfy the diffraction conditions (see Chap.2).[2]

---

[1] For a given electronic configuration of an atom, there are different possible distributions of the electrons in the atomic orbitals, which implies several possible total angular (orbital) and spin momentums. For each pair of orbital and spin momentums, we obtain a state and the one with lowest energy is the ground state. However, each state may be produced by several microstates, depending on the exact distribution of electrons in each orbital type. A calculation normally refers to one such microstates and therefore may return a non-spherical electron distribution around an atom, whereas an average among microstates necessarily returns a spherical electron density.

[2] Noteworthy, sometime the quantity $\mathbf{H}$ is used to indicate the vector in the reciprocal space which satisfies diffraction conditions, where $|\mathbf{H}| = 1/d_{\mathbf{H}} = |\mathbf{k}_{diffracted} - \mathbf{k}_{incident}|/2\pi = 2sin\vartheta/\lambda$, while $\mathbf{k}\ or\ \mathbf{S}$ (with $\mathbf{k} = 2\pi\mathbf{S}$) are used to address a generic point in the reciprocal space. Therefore, the

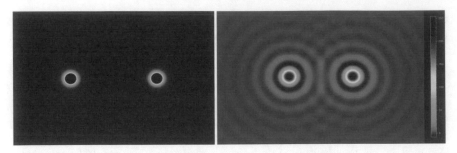

Fig. 3.4 **Left**: The electron density of the O_2 molecule in direct position space from theoretical calculations. **Right**: The electron density of the O_2 molecule obtained from the Fourier transform of the form factors calculated from the very same electron density plotted on the left, truncated at $sin\vartheta/\lambda = 1.9\text{Å}^{-1}$. Both functions are plotted as bitmaps with a color-coded scale represented on the right. The important features are the ripples generated by the truncation. Noteworthy, this resolution is higher than what ideally available with Ag Kα radiation

A few comments on Eq. (3.4):

- The Fourier transform is a series instead of an integral because the points bearing a non-zero amplitude are discretely distributed in the reciprocal space, in virtue of the Laue conditions, whereas the scattering matter (the electron density) in direct space is continuously distributed.
- The $1/V$ in the equation simply represents the reciprocal space volume associated to each node point $\mathbf{k}$ and it corresponds to the inverse of unit cell volume in the direct space.
- The structure factor $F_{\mathbf{k}}$ is a complex number, thus one should write it as $F_{\mathbf{k}}e^{i\varphi_{\mathbf{k}}}$, where $F_{\mathbf{k}}$ is the modulus and $\varphi_{\mathbf{k}}$ is the phase of the reflection $\mathbf{k}$. The phase depends on the distance between the sources of the waves that interfere to produce the scattered intensity. As anticipated in Chap. 2, because the intensity of the wave is proportional to $|F_{\mathbf{k}}|^2$, but independent from $e^{i\varphi_{\mathbf{k}}}$, the measured diffracted intensity does not provide the full information necessary to calculate $\rho(\mathbf{r})$.
- The series is virtually infinite, which means that for an exact reconstruction of the scattering matter, one needs an infinite number of structure factors amplitudes. This is impossible in practice, because the finite wavelength λ necessarily defines the number of nodes of the reciprocal space that can be measured and that are included in the summation (3.4).

These considerations lead immediately to the conclusion that Eq. (3.4) cannot be the way in which one can map the electron density accurately. In fact, the maps obtained in that way, even after solving the problem of not knowing $\varphi_{\mathbf{k}}$, would be inadequate, because affected by ripples caused by the series termination. An example is shown in Fig. 3.4, where for the oxygen molecule the electron density in the direct space (from a quantum mechanical calculation) is compared to what one would obtain

---

coordinates of **H** in the reciprocal space are integer numbers h, k, l, those emerging from the Laue conditions.

after applying a Fourier transform and then a back transform omitting structure factors beyond a given resolution limit. Noteworthy, in this virtual experiment we have not considered effects due to atomic motion or experimental errors.

Thus, prior information is necessary to make the picture more stable and more intelligible. There are several ways to reach this goal. In this chapter, we will analyze methods based solely on the electron density, whereas in Chap. 5, we will discuss wavefunction based methods.

3.2 Experimental Deformation Density

The first of the approaches to circumvent the series termination problem of Eq. (3.4) is through cancellation of the ripples and enhancement of the deviations from atomic sphericity. This implies visualizing an experimental deformation density, which can be obtained using Eq. (3.4), but subtracting from the exact structure factors those computed with a model density, like the promolecule density. This is a representation of how much the true electron density $\rho(\mathbf{r})$ deviates from that calculated with the IAM model. Because both $F_{\mathbf{k},expt}$ and $F_{\mathbf{k},IAM}$ are truncated at the same resolution limit, the ripples are mutually cancelled, and the result is a quite good approximation of the exact deformation density $\Delta\rho(\mathbf{r})$ (see for example Verschoor 1964; and Coppens 1967):

$$\Delta\rho(\mathbf{r}) = \rho(\mathbf{r}) - \rho_{IAM}(\mathbf{r}) \sim \frac{1}{V_{unit\ cell}} \sum_{\mathbf{k}} \left[\left(F_{\mathbf{k},expt} - F_{\mathbf{k},IAM} \right) e^{i\varphi_{\mathbf{k},IAM}} \right] e^{-i\mathbf{kr}} \quad (3.5)$$

In Eq. (3.5), the phase of each measured reflection is approximated with that calculated with the IAM model. In fact, while the experimental phase is unknown, if IAM is so-far the best available model, then $\varphi_{\mathbf{k},IAM}$ is the best estimation of $\varphi_{\mathbf{k}}$. The outcome of the summation (3.5) is a difference density map, which addresses the accumulation or depletion of electron density at each point $\mathbf{r}$ with respect to the IAM, see an example in Fig. 3.5 for the molecule of benzene. In this plot, instead of using experimentally measured structure factors, $F_{\mathbf{k},theor}$ are used. They are computed from the electron density obtained in direct space by means of a theoretical calculation. This allows us to compare the effect of the transformation of $\rho(\mathbf{r}) - \rho_{IAM}(\mathbf{r})$. At variance from Fig. 3.4 (where we calculate the total density), the result in Fig. 3.5 (calculating the deformation density) is more satisfactory.

The integral of $\Delta\rho(\mathbf{r})$ over volume is zero, whereas the integral of the electron density from Eqs. (3.3) or (3.4) return the total number of electrons in the unit cell:

$$\int_{unit\ cell} \Delta\rho(\mathbf{r})d^3r = \int_{unit\ cell} [\rho(\mathbf{r}) - \rho_{IAM}(\mathbf{r})]d^3r = N_{el} - N_{el} = 0 \quad (3.6)$$

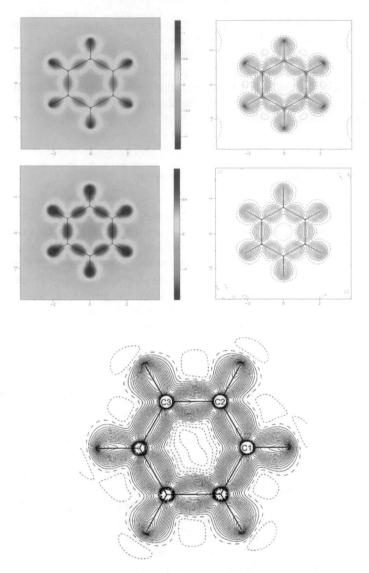

Fig. 3.5 The deformation density in the benzene molecule **Top**: The calculated deformation density from a theoretical electron density calculation. **Center**: The deformation density obtained via Fourier transformation of the structure factors computed from the very same density used for the top picture. For both functions, plots on the right are color-coded bitmaps, those on the right are contour plots (positive contours are blue solid lines, negative contours are red dotted lines, and zero contour is a black dashed line). **Bottom**: The experimental static deformation density of benzene reported by Bürgi et al. (2002) based on the multipolar model refined against data collected at 110 K (picture reproduced with permission of John Wiley and Sons)

It would be obvious to conclude that the deformation density straightforwardly reveals the differences between isolated atoms and chemically bonded atoms. As a matter of fact, Fig. 3.5 seamlessly reveal the chemical bonds. However, this interpretation is not entirely correct: the deformation density reveals the difference between an isolated atom and an atom inserted into an electric field, like the crystal field (in general) and in particular the molecular field if the atom is covalently bonded to others. Because the field in a crystal cannot be spherically symmetric, the atomic electron density, even in the absence of covalent bonds, is necessarily aspherically perturbed, in keeping with the crystallographic site symmetry. In Fig. 3.6, we see the effect of an electric field applied to an isolated oxygen atom and again we can spot a significant and visible difference between the atomic form factor in isolation, the form factor in a direction parallel and perpendicular to the applied field.

Noteworthy, although usually assumed to be spherical, atomic ions may also manifest aspherical deviations. This is well known for transition metals involved in complexes through coordinative bonds with organic or inorganic ligands. The deformation depends on the uneven occupation of their $(n - 1)d$ orbitals induced by the ligand field. Originally the field was called in fact crystal field, assumed to be generated only by the electrostatic interactions with the environment (crystalline because most of these compounds were studied as crystals). The term *crystal field theory* (Bethe 1929) was typically adopted to explain the origin of uneven occupation of metal *d*-orbitals, although the *ligand field theory* (Van Vleck 1932; Figgis 1966) became afterwards more popular, because it assumes that the metal deformation depends also on an orbital interaction between the ligands and the metal ions.

This is the reason why the deformation density of transition metal atoms or ions in these complexes is so pronounced, compared to that of alkali metal ions in salts lacking significant overlap intervention (see for example the tiny deviations

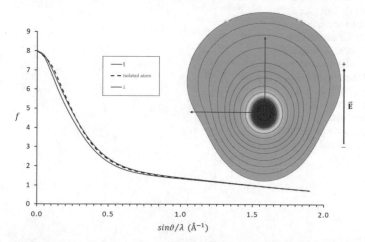

Fig. 3.6 The electron density of an oxygen atom in an electric field and the corresponding form factor in a direction parallel or perpendicular to the field. As a reference the form factor of the unperturbed atom is also plotted

of observed for fluorite by Kurki-Suonio and Meislao 1966). In Fig. 3.7, one can see the deformation density around Cr (with a formal oxidation state of 0) in $Cr(CO)_6$, a prototypical metal carbonyl complex, featuring a pseudo-octahedral symmetry in the crystal (close to the exact octahedral symmetry that occurs *in vacuo*).

Let us go back to the interpretation of deformation densities in more classical covalent bonds. From the question "can we see the electrons?", we now ask: *can we see the chemical bonds*? A big warning is necessary at this point. Quantum mechanics tells us what observables are (see above) and what is the expectation value of an observable (the average value of the operator acting on the wavefunction). However, there is no operator associated with the "chemical bond", which is a pure abstraction, though extremely useful to explain and describe compounds. The definition itself of a chemical bond is elusive. An example is the one provided by Pauling (1939):

Fig. 3.7 The deformation density in a plane of the pseudo-octahedral $Cr(CO)_6$. **Top**: static deformation density recalculated from the multipolar refinement of X-ray diffraction experiments by Farrugia and Evans (2005). The plane contains four carbonyls and the metal. One can clearly see the electron density depletions at the central Cr atom, due to the field produced by the CO ligands that attack the metal with their available lone pair at C atoms. Noteworthy, the electrons at Cr accumulate out of the plane. **Bottom**: the experimental deformation map published by Jost, Rees and Yelon (1975) by combining neutron and X-ray diffraction (reproduced with permission of the International Union of Crystallography)

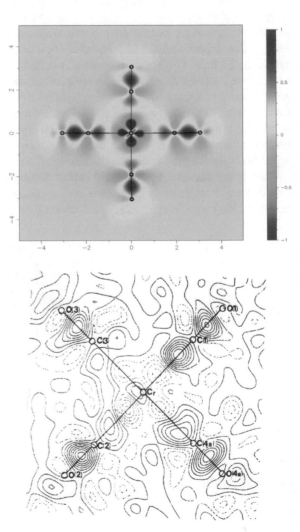

We shall say that there is a chemical bond between two atoms or group of atoms in case that the forces acting between them are such as to lead to the formation of an aggregate with sufficient stability to make it convenient for the chemist to consider it as an independent molecular species.

Note that there is no *observable* directly invoked in this definition and the reference to a convenient situation for the chemist is quite "horrible" in physical terms. A more recent definition, as that approved by IUPAC (Minkin 1991), tries some correlation with observables, though without any sharp cutoff:

When forces acting between two atoms or groups of atoms lead to the formation of a stable independent molecular entity, a chemical bond is considered to exist between these atoms or groups. The principal characteristic of a bond in a molecule is the existence of a region between the nuclei of constant potential contours that allows the potential energy to improve substantially by atomic contraction at the expense of only a small increase in kinetic energy. Not only directed covalent bonds characteristic of organic compounds, but also bonds such as those existing between sodium cations and chloride anions in a crystal of sodium chloride or the bonds binding aluminium to six molecules of water in its environment, and even weak bonds that link two molecules of O_2 into O_4, are to be attributed to chemical bonds.

We may conclude that the question "can we see the bonds?" is improper. We cannot see something that is not *per se* quantum mechanically observable, and this unsuccess does not depend on the failure of an experiment or more generally on the inability of a technique. We simply *cannot* observe chemical bonds. Nevertheless, we can see many effects and consequences of what we call chemical bond. This is quite common in Chemistry, because we typically identify, characterize, and use molecules that are built from atoms through chemical bonds. The experimental evidence that molecules exist depends on their constituting atoms being bonded. Obviously, we dare being a bit more precise in Quantum Crystallography and characterize finer details that not only reveal the bonds, but also their nature, their consistency, their role in determining a given physical property, etc.

One may be tempted to associate the existence of a chemical bond with the accumulation of electron density in the region in between the two atoms, because we consider electrons as the "glue" that keep nuclei close to each other. This was undoubtedly the reason of the success of the early experimental works on electron density determination (Coppens 1967). Although this assumption is reasonable, it is not fully correct.

Let's see why.

First, we should clarify the meaning of two different concepts: *binding* and *bonding*. Using the words of Berlin (1951), we can address these two distinct aspects of molecule formation:

The word "binding" shall be defined in such a way that it relates to the forces acting on the nuclei in the molecule. Thus, binding shall be distinguished from bonding which is usually related to the energy of the molecule. The bonding by a single electron is related to the energy of this electron in the molecule; in the same sense, the binding by a single electron will be related to the forces exerted by this electron on the nuclei.

Thus, *binding* concerns the (attractive) electric forces, whereas *bonding* concerns the (stabilizing) energy of electrons. Although the chemical bond is more genuinely correlated with the *binding* ("forces acting on atoms"), in the definitions there is a promiscuous mix up with energy ("stabilization of an aggregate") and usually the latter is used to infer the presence of a bond. As a matter of facts, the *strength* of a chemical bond is often ascertained with the bond enthalpy instead of its actual force constant.

Berlin's approach was assessing for each point in the position space if the electric force, generated by the electronic charge at the point, induces an attraction or otherwise a repulsion between two nuclei. If we calculate the electric field vector **E** generated by the nuclear charges at the point **r**, and project it along the internuclear vector, we see whether a negative electric charge positioned at **r** would contribute to compressing or stretching the internuclear distance. Due to the Hellmann-Feynman theorem (Hellmann 1937; Feynman 1939), the sum of all these contributions (thus the force generated by all charges distributed in space) must be zero for a system in equilibrium. The deformation density, therefore, should be considered within this paradigm: accumulation of charge in some region contributes to the binding, whereas in other region it does not. Obviously, the binding regions are those in between the nuclei, asymptotically delimited by directions inclined 45° with respect to the internuclear vector for the simple case of homopolar diatomic molecules (see Fig. 3.8). In keeping with this interpretation, Spackman and Maslen (1985) suggested to apply Berlin's criteria to the deformation density rather than to the total electron density. This would emphasize the idea that it is the electron reorganization with respect to the isolated electronic ground states of atoms which is the most evident manifestation of chemical bonding or otherwise the evidence of a missing chemical bond.

Fig. 3.8 The Binding/Antibinding regions of a diatomic homopolar molecule. Image reproduced from Berlin (1951) with permission of AIP-Publishing

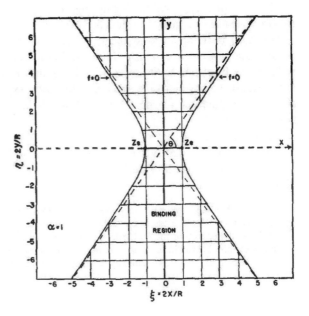

This is very simple to understand, but there is a serious concern. Once aggregated with others, atoms do not know what their ground electronic state in isolation is. Arguably, the sum of isolated electron densities is an ambiguous reference to measure the *mise en place* of a chemical bond. While for many covalent bonds in organic molecules the deformation density does in fact reveal charge accumulation in the internuclear region, thus corroborating the above-mentioned hypothesis, some problematic cases are for example single bonds between atoms of the late second period, like F-F and (to a smaller extent) O-O.

The reason is quite simple. The spherically averaged electron density of fluorine in its electronic ground state is the mean between some microstates. In fact, the atomic ground state term ^2P of fluorine, derived from the electronic configuration $1s^2 2s^2 2p^5$, is sixfold degenerate: the singly occupied p orbital could be p_x, p_y, or p_z, each hosting a spin α or β. In the absence of magnetic fields, and ignoring spin-orbit coupling, the 6 microstates are equivalent (degenerate). The average implies that along the direction of the F-F bond (assumed as z), the p_z orbital of the reference state has a nominal occupancy of 5/3 e. However, in the ideal situation of forming a 2-center-2-electron bond with the other F atom, p_z is supposed to host only one electron to form a doubly occupied molecular orbital through a linear combination with the p_z orbital of the other F atom, also singly occupied. It appears quite clear that the reference promolecule is electron richer compared with the true molecule and, accordingly, a depletion of electron density is found, instead of an accumulation, despite the rather obvious bonding situation. While this is easily explained for such a simple biatomic molecule, for more complicated molecules it may be not so obvious. For example, a long debate started concerning the missing accumulation of electron density in molecules containing bonds between transition metals, some of them were also measured experimentally with X-ray diffraction, like $Mn_2(CO)_{10}$ (Martin et al. 1982) or $Co_2(CO)_8$ (Leung and Coppens 1983). An additional complication comes from the large size of the metal atoms, which makes the binding region much broader and therefore an even smaller charge accumulation may produce substantial attractive force, albeit being less visible.

A solution for the reference problem considers the particular microstate of interest, which for F_2 correspond to the occupancy $1s^2 2s^2 2p_x^2 2p_y^2 2p_z^1$. See Fig. 3.9 for a graphical representation.

In modern crystallography, the covalent intramolecular bonds are not the most important ones, because it is the interaction occurring between molecules to be central, even more for crystal engineering. For weaker interactions, the deformation density is also problematic. Again, the longer internuclear distance makes the binding region quite large and the low covalency of intermolecular interactions does not reflect in substantial electron density accumulation between atoms. Only very strong hydrogen bonds exceptionally produce substantial reorganization especially at the acceptor atom to be quite visible with sufficiently accurate measurements.

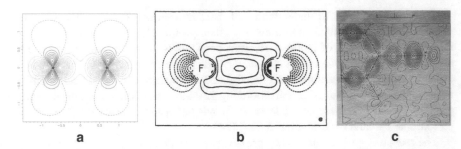

Fig. 3.9   a The deformation density of F_2 computed by subtracting the *promolecule* electron density from the electron density of the molecule. Blue contours represent electron density accumulation, whereas red contours reveal electron density depletion with respect to the reference. Because the spherically averaged electron density of F in its atomic ground state implies having 5/3 electrons in each p orbital, the amount of promolecule electron density subtracted along the bond direction (2 x 5/3 e = 10/3 e) largely exceeds the number of electrons expected for the formation of a F-F single bond (2 e). As a result, one obtains a "surprising" depletion of electron density in the bond region. **b** The deformation density calculated by Kunze and Hall (1986) using valence state sp hybridized atoms as reference instead of the spherical atoms as in the top picture. Reproduced with the permission of the American Chemical Society. **c** Deformation density of tetrafluoroterephtalo-dinitrile where the C-F bond is compared with C-C and C-N bonds, showing the lack of charge accumulation in C-F bonds (reproduced with permission of John Wiley and Son from Dunitz et al. 1983)

3.3   Models for the Total Electron Charge Density

A second way to circumvent the problem of mapping the electron density in a crystal is to refine a model of the total electron density and obtain therefore an analytical function. This is conceptually not different from the routine crystal structure determination, which returns a crude electron density model (IAM) based on the refined atomic positions and displacements from equilibrium. An atom-based accurate density model is a step forward from the IAM.

3.3.1   Modelling the Static Electron Density: The Multipolar Model

In the early 1970s, several scientists approached the problem and proposed recipes, sharing the idea that the atomic electron densities can be anyway treated "independently" like for the IAM. Thus, the total electron density is still a sum of atomic contributions, but the atoms are now not neutral and not spherical. The physical grounds of this model came from four seminal papers by McWeneey (1951, 1952, 1953, 1954) who proposed aspherical atomic form factors for a better modelling of the bonding electron density in crystals. The formalism was later revised and gener-alized by Dawson (1964, 1967), Stewart (1969), and Stewart et al. (1975), based on the concept of the so-called *generalized atomic form factors*, obtained as Fourier transform of the molecular electron density projected onto atom centers (for the case of molecular crystals).

Here it comes the needs to define an atom in a context, which is different from that of the pure IAM. Stewart proposed the so-called *pseudoatom*, an atom defined by a series of functions representing the projection of the total (molecular/crystal) electron density onto it. Thus, the model is still atom centered, but no longer *independent* in terms of electron density contributions because the atomic electron densities are mutually correlated in this approach and they are not predetermined. On the other hand, the atomic displacements remain independent from each other, like in IAM. The result is a new electronic form factor (different from IAM) for each atom and associated with a Debye-Waller factor that describes the independent atomic displacements from their equilibrium positions (just like for IAM).

Stewart (1976) also proposed the refinement of the coefficients of the pseudoatom from the measured structure factors using the least-squares method, which is normally adopted for IAM refinements as well. Other authors proposed similar models (for example, Kurki-Suonio 1968; Hirshfeld 1971), but the most adopted one was eventually that devised by Hansen and Coppens (1978), who described the aspherical atomic density $\rho_{i,asph}$ with the following expansion (see Fig. 3.10 for the molecule of H_2 and the individual contribution of one H atom to the molecular density):

$$\rho_{i,asph}(\mathbf{r} - \mathbf{r}_0) = \rho_{i,core}(\mathbf{r} - \mathbf{r}_0) + P_{i,valence}\kappa_i^3 \rho_{i,valence}(\kappa_i(\mathbf{r} - \mathbf{r}_0))$$

$$+ \sum_{l=0,}^{l_{max}} \left[\kappa_i'^3 R_{i,l}\left(\kappa_i'(\mathbf{r} - \mathbf{r}_0)\right) \sum_{m=0,l} P_{i,lm\pm} y_{lm\pm}((\mathbf{r} - \mathbf{r}_0)/r) \right] \quad (3.7)$$

In this formalism, the parameters to refine, by means of least squares minimization, are: the coordinates of the atomic nucleus ($\mathbf{r}_0$); the displacement parameters of each atom i (part of the Debye Waller factors like for the IAM refinement, see Eq. 2.4); the population coefficients of the atomic multipoles ($P_{i,valence}$, $P_{i,lm\pm}$); other parameters describing the contraction or expansion of the atomic valence shells (κ_i^3, $\kappa_i'^3$). $\rho_{i,core}(\mathbf{r})$ and $\rho_{i,valence}(\mathbf{r})$ are spherical electron density functions representing the core and valence shells and constructed from the pertinent atomic orbitals. The latter come from quantum mechanical atomic wavefunction calculations of the electronic ground state of the isolated atom, thus identical to the situation of IAM. However, the population parameters introduced for the spherical valence density is a deviation from IAM because each atom is allowed to increase or decrease the number of valence electrons and resulting no longer neutral. For the core density, the population is typically frozen to the formal one. Some model extensions also enable refinement of these parameters that typically deviate only marginally from the formal number (Bentley and Stewart 1974; Fischer et al. 2011). The last part of Eq. (3.7) is the most substantial difference compared with IAM, because it contains the expansion in terms of multipoles exceeding the spherical monopole and including dipoles, quadrupoles, octupoles, etc. These aspherical distributions are described through

Fig. 3.10 **Top**: the
multipole-reconstructed
electron density in the H_2
molecule. **Bottom**: the
multipolar electron density
of one H atom in the H_2
molecule. The top picture is
simply the summation of the
two multipolar expansions of
each H atom

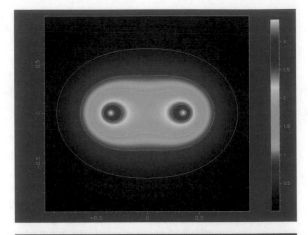

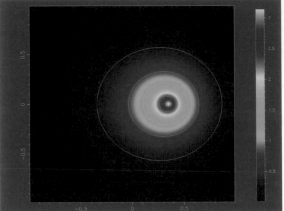

spherical harmonics ($y_{lm\pm}$), the set of orthogonal functions used to solve differential equations. They are not unknown in quantum mechanics because they do in fact constitute the angular part of the solutions of the Schrödinger equation for hydrogen-like atoms (mono-electronic atoms like hydrogen but with all possible nuclear charges). They indeed represent the shape of what we normally call *orbitals*.

Why are spherical harmonics useful for the electron density as well and what is their meaning here? The answer is easy. From the second postulate of quantum mechanics, we know that a wavefunction is a probability amplitude and that the probability is proportional to the square of the modulus of the wavefunction. Thus, the electron density of an atomic orbital is proportional to the square of the orbital, which means the square of its radial and of its angular part. While the former is easy, the second is apparently more complicated. However, the spherical harmonics do possess an important property, the *closure*. This means that the product of two spherical harmonics is a linear combination of spherical harmonics. This is obvious for a monopole (a spherical function whose square is again a spherical function) and the same holds true for all other spherical harmonics. Thus, this set of function is the best to describe not only the angular part of the atomic orbitals but also the angular part of the atomic orbital density, onto which we project the electron density of the whole system according to the pseudoatom formalism.

The type of atomic valence orbitals does influence the expansion, in fact the product of spherical harmonics $y_{l_1,m_{l_1}} \times y_{l_2,m_{l_2}}$ always produces terms that with l running from $|l_1 - l_2|$ to $l_1 + l_2$. This already implies that atoms having p-orbitals ($l = 1$) in their valence need at least *quadrupolar* terms in the expansion ($l = 2$) and transition metals with d-orbitals in their (inner) valence do require at least *hexadecapoles* ($l = 4$) and f-orbitals ($l = 3$) would require at least *hexacontatetrapole* ($l = 6$). Moreover, the presence of strong overlap among atomic orbitals, as for covalent bonds involving second period atoms, do require an even larger expansion to capture the electron density due to the two-center orbitalic products. In fact, an expansion to at least octupoles is necessary for atoms of the second period, especially to describe the trigonal-like densities occurring in sp^2 hybridized atoms. Hexadecapoles could be necessary for sp^3 hybridization (and pseudo-tetrahedral stereochemistry). See more details in McWeeney (1954) and Dos Santos et al. (2014).

The radial density part $R_{i,l}(\kappa'_l \mathbf{r})$ is again borrowed from atomic wavefunction calculations and it coincides with the square of the radial functions calculated for valence orbitals of electronic ground states for isolated atoms (Clementi and Raimondi 1968).

From Eq. (3.7), we realize the much larger number of parameters necessary for a multipolar model, which obviously implies some possible problems, especially the correlation among them during a least square refinement. One should remember that the number of coefficients necessary for the last term of Eq. (3.7) is $n = (l_{max} + 1)^2$ where l_{max} is the highest order of spherical harmonics employed for the multipolar expansion ($l = 0, 1, 2, \ldots l_{max}$). Typically, $l_{max} = 4$, which implies 25 parameters to refine, in addition to the three coordinates of the atomic nuclear position and 6 parameters of the anisotropic displacement of the atom from the equilibrium position. Thus, for each atom in the asymmetric unit, a typical number of parameters – in the absence of symmetry constraints – is 34, instead of 9 as in conventional IAM refinements. Moreover, the contraction/expansion parameters κ_i and κ'_i are variables, although typically set identical for all atoms of each atom type in the structure to avoid divergence of the refinement.

Because of this explosion of parameters, a multipolar model requires going well beyond the typical resolution achieved for routine crystal structure determinations. Crystallographers often consider a reasonable limit for the refinement of a standard structural model the so-called *Cu-sphere*, i.e. the maximal resolution obtainable using the wavelength of Cu K α radiation ($d = 0.77$ Å), and the International Union of Crystallography (IUCr) recommends a resolution of $d = 0.83$ Å. This limit is safe enough for routine crystal structure determinations but would not suffice to exhaust the increased complexity of a model like Hansen and Coppens'. It is instead necessary achieving at least $d = 0.50$ Å or below (i.e. $sin\vartheta/\lambda \geq 1.0$ Å^{-1}), to have enough data to refine a multipolar model keeping the rule of thumb that the reflections/parameters ratio should be in excess of 10. Things are not so easy, though. In fact, the scattering of valence shell electrons of a second period atom decays very rapidly with the diffraction angle (see Fig. 3.2). Thus, the vast majority of the reflections are unusable for refining the multipole coefficients. Roversi et al. (1996) discussed the problem in terms of "efficient" reflections and pointed out that the reflection/parameter ratio is inherently well below 10, a serious issue causing the undesired correlation. All the outer reflections, with contribution almost entirely from core electrons, inform only on the position and displacement of atoms.

Despite these possible pitfalls, the multipolar model proved over the years to be extremely useful and enabled the experimental determination of the electron density distribution in a number of molecular crystals, as well as of polymeric and inorganic samples. Scientists have developed many computer programs, adopting the different variants of multipole models, and the most diffused of them are listed in Table 3.1.

The multipolar model represents the first successful attempt to extract an accurate and reliable static electron density from X-ray diffraction data that necessarily contains also the effects of lattice dynamics. Atoms are of course not steady, as anticipated and their oscillation is the main reason of an attenuation of their scattering power. This has an implication for the measurement itself, which needs to be longer due to the loss of intensity at Bragg diffraction positions in the reciprocal space. Moreover, the reduced scattering significantly affects the successful deconvolution of features that are genuinely due to the electron density from artifacts caused by the atomic motion. It is well known that the lattice vibrations affect the structural models and that large atomic displacement from their equilibrium positions make interatomic distances apparently shorter (Schomaker and Trueblood 1968). The problem is even more important if the purpose is mapping the electron density distribution because the electron density bound to a nucleus moves with it and genuine features of the chemical bonds are mixed with those due to the atomic displacements. While one cannot remove the atomic motion completely, at low temperature it is at least significantly reduced. This partially solves both problems: the accurate positioning of atoms and the accurate description of the electron density, especially of its deviations from sphericity. In Fig. 3.11, we see the effect on the atomic form factor due to the atomic displacement (through the Debye Waller factor). As it can be see, the form factor is particularly affected at high angle. In a seminal paper, F. L. Hirshfeld (1976) tried to answer the question "Can X-ray data distinguish bonding effects from vibrational smearing?". Hirshfeld proposed to adopt a criterion to access the conditions at which we may be confident: the difference between atomic displacement parameters projected along bond distance vectors should not exceed a (low enough) limit. Below this limit, which was empirically chosen, one may suppose that a rigid body motion of the molecule occurs and therefore separation of electron density deformations from atomic motion is doable. This analysis took the name of Hirshfeld rigid bond test. It is not universally valid, because it was tuned mainly for organic molecules and relatively low mass atoms (C, N, O), but it provides useful clues of the feasibility of a quantum crystallographic study (true not only for the multipolar model refinement).

Hirshfeld (1977a) explicitly stated that a perfect deconvolution of the static electron density distribution, complying with the Born-Oppenheimer approximation, is impossible. However, the combination of a suitable electron density model, like

Table 3.1 A list of the most common programs employed for multipolar expansion (refinement and/or calculation of the properties)

Authors	Program name	Program type	Multipolar model
Hirshfeld (1971)	**LSEXP**	Least square refinement of deformation density coefficients	Hirshfeld (1971)
Hansen and Coppens (1978)	**MOLLY**	Least square refinement of atomic multipolar expansion	Hansen and Coppens (1978)
Figgis et al. (1980)	**ASRED**	Least square refinement of Valence orbital multipolar model	Figgis et al. (1980)
Stewart and Spackman (1983), Stewart et al. (2000)	**VALRAY, VALRAY2000**	Least square refinement of atomic multipolar expansion and calculation of density-based properties	Stewart (1976)
Craven et al. (1987)	**POP**	Least square refinement of atomic multipolar expansion	Stewart (1976)
Koritsanszky et al. (1995), Volkov et al. (2006, 2016)	**XD, XD2006, XD2016**	Least square refinement of atomic multipolar expansion and calculation of density-based properties	Hansen and Coppens (1978)
Stash and Tsirelson (2002, 2014)	**WINXPRO**	Calculation of density-based properties from multipolar models	Hansen and Coppens (1978)
Jelsch et al. (2005)	**MOPRO**	Least square refinement of atomic multipolar expansion and calculation of density-based properties	Hansen and Coppens (1978)
Deutsch et al. (2012)	**MOLLY-XN**	Least square refinement of spin-resolved atomic multipolar expansions	Hansen and Coppens (1978), Deutsch et al. (2012)
Petricek et al. (2014)	**JANA2006**	Least square refinement of atomic multipolar expansion	Hansen and Coppens (1978)

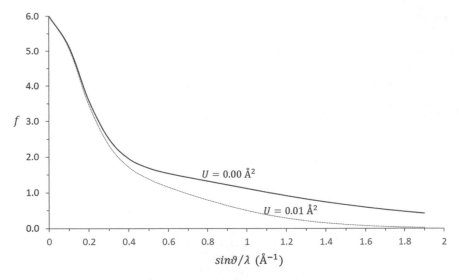

Fig. 3.11 The atomic form factor of C atom with and without displacement of the nucleus from equilibrium position. The mean atomic displacement is indicated with U

the multipolar model, and low temperature measurements may suffice to extract a suitable electron density, comparable with that calculated for a given static geometry.

3.3.2 Modelling the Dynamic Electron Density

Although the de-smeared electron density is normally a goal of the quantum crystallographic research, dynamic electron density distributions have also been computed and analysed. This is technically possible, but more complicated, with a direct space analytical model of the electron density, like the multipolar model. The reason is that the dynamic density is the convolution of the static density and the nuclear probability distribution function. Thus, other approaches are preferable, that reconstruct a total dynamic electron density $\rho_{dyn}(\mathbf{r})$. While a Fourier synthesis (Eq. 3.4) would suffer of the above-mentioned ripples, the best solution is adopting the maximum entropy method (MEM). This method has the advantage of optimizing the quality of the total function of a given signal. It was not introduced to specifically solve problems of crystallography, but it has been widely adopted also for solving or refining crystal structures, especially from very noisy or complicated data, like powder diffraction.

In the 1990's, MEM was proposed also for modelling accurate electron density from X-ray diffraction (see for example, Takata and Sakata 1996). MEM is based on the information entropy S (Jaynes 1968):

$$S = -\sum_{j=1}^{N_p} \rho_j(\mathbf{r}_j) ln\left(\frac{\rho_j(\mathbf{r}_j)}{\rho_j^{prior}(\mathbf{r}_j)}\right) \qquad (3.8)$$

where N_p is the total number of points $\mathbf{r}_j$ in a grid of the unit cell, $\rho(\mathbf{r}_j)$ is the electron density at the point $\mathbf{r}_j$ and $\rho_j^{prior}(\mathbf{r}_j)$ is the corresponding value for a convenient but ideally non-biased *prior* reference electron density. Maximizing S means finding the most likely distribution of $\rho_j(\mathbf{r}_j)$, which is therefore not analytically reconstructed but sampled over the N_p points in a unit cell. $\rho_j^{prior}(\mathbf{r}_j)$ may play a fundamental role: it could be taken as a uniform function or otherwise a non-uniform one, perhaps starting from a preliminary model, like IAM, or even an advanced model like a multipolar model. It is interesting that in the first case, there is no quantum mechanical bias in the preliminary information, whereas in the non-uniform prior there could be a significant bias. Roversi et al. (1998) investigated in details the various aspects and suggested that MEM could be even used to further improve the multipolar model. Other studies also addressed the effects of choosing the prior-density, see for example Palatinus and van Smaalen (2005) or van Smaalen and Netzel (2009), or compared the MEM and multipolar model approaches (Hoffmann et al. 2007).

The analysis of the reconstructed electron density is of course more complicated to compare with theoretical predictions, which are usually done on static densities calculated with quantum chemical methods. Nevertheless, there are some seminal studies which provided unprecedented information, like the analysis of electron density in metallic Be (Iversen et al. 1995, see Fig. 3.12) or in hydrogen bonded systems (Netzel and van Smaalen 2009). More recently, Hübschle and van Smaalen (2017) used MEM reconstructed electron densities to calculate also a dynamic electrostatic potential and proposed an in-depth analysis.

3.4 Models for the Electron Spin Density

3.4.1 The Spin Density and Experimental Techniques

Although many molecular crystals contain closed-shell molecules, interacting through electrostatic and London-type forces (Eisenschitz and London 1930), there is an emerging interest for crystals consisting of open shell molecules or polymers, in which there is an excess of electrons with one type of spin over the other. As a matter of facts, many extended solids contain spin active metal ions. Moreover, organic radicals, metal-complexes or coordination polymers form crystalline materials that are attracting increasing attention.

In the field of *spintronics*, the coupling and ordering of magnetic moments are fundamental characteristics. The simpler macroscopic magnetism is described with the collinear assembly of atomic moments, giving rise to parallel alignment

Fig. 3.12 The electron density in a plane of Be crystal calculated with MEM. Reproduced from Iversen et al. (1995) with permission of the International Union of Crystallography

(*ferromagnetic*), antiparallel alignment of identical moments (*antiferromagnetic*) or antiparallel alignment of non-identical moments leaving a residual magnetic moment (*ferrimagnetism*). More intriguing coupling is that between non collinear moments, for example the *spiral* systems. The current research focuses on new species featuring elusive types of magnetic interactions that could favor quantum information processes. A huge array of experiments and observations provide information for a comprehensive quantum mechanical characterization of a magnetic system, for example superconductive quantum interference, muon spectroscopy, nuclear magnetic resonance, electron spin resonance, etc. Most of these are feasible on crystalline species, giving insight in the magnetism of the species and therefore being very useful for the reconstruction of the quantum mechanical functions, especially the spin electron density, $\sigma(\mathbf{r})$.

The spin density is, like the charge density, a probability distribution (defined in position or in momentum space) of an intrinsic property carried by electrons. More precisely, the spin density is the difference between the probability of finding one electron with spin α and that of finding one electron with spin β.

$$\sigma(\mathbf{r}) = \rho^{\alpha}(\mathbf{r}) - \rho^{\beta}(\mathbf{r}) \tag{3.9}$$

It is important that the electron spin is a source, but not the only one, of the magnetization density. The magnetization depends both on the *spin angular momentum S* and on the *orbital angular momentum L*. The latter is definable also in a classical physics atomic model (albeit incorrectly), whereas the spin is inherently quantum mechanical. Gerlach and Stern (1922) discovered its existence experimentally, while trying to prove the quantization of the angular momentum. To explain the experiment, Uhlenbeck and Goudsmit (1925) hypothesized the presence of an additional angular momentum for electrons, namely the spin. Thus, all research on spin density determination in crystals, and in general about magnetization, is inherently quantum crystallographic. Here, we will mainly focus on the determination and the analysis of the spin density.

The best way to observe the spin density is through a radiation carrying itself a spin, like neutrons. The neutron elastic scattering from crystals is governed by the same Bragg law valid for X-rays, despite the different nature of the radiation and the different atomic cross section. The measured intensities depend on two different kinds of interactions of the radiation: (a) with the atomic nuclei (giving rise to the nuclear structure factor F_N) and (b) with the atomic magnetic moments (F_M, if any).

In charge density analysis, the nuclear structure factor is useful to model very accurately the position and thermal motion of atoms, especially of hydrogens that are problematic with X-rays. In the past, this allowed an independent structural model with respect to that obtained from X-ray diffraction, and therefore free from biases due to the bonding electron density. Thus, the combination of X-ray and neutron scattering enabled high quality density maps, although this technique is now less frequently used. For the spin density distribution, instead, one should focus on the magnetic part of the neutron scattering. In analogy with the charge density, one can reconstruct the magnetization density from the magnetic scattering factors measured at Bragg positions:

$$\mathbf{m(r)} = \frac{1}{V} \sum_{\mathbf{H}} \mathbf{F}_M(\mathbf{H}) e^{2\pi i \mathbf{Hr}} \qquad (3.10)$$

The summation must also include the $\mathbf{H} = 0$ term, which represents the total magnetization of a unit cell. The magnetic form factor measures the projection of the magnetization onto a plane perpendicular to $\mathbf{H}$ and therefore there cannot be a scattering when the magnetic moment and the scattering vector are parallel. This implies that the measurement of the magnetization density is inherently ambiguous, in keeping with the lack of determination for the angular momentum. However, the spin density is not ambiguous, because it determines an excess of spin population and its distribution, but not the direction of the spin magnetic moment.

For heavier elements, the treatment of magnetization density cannot exclude the spin-orbital coupling, leading to a more complicated magnetization density.

Using a polarized neutron beam, one can measure the so-called *flipping ratios* (see Brown 1992, Matthewman et al. 1982), i.e. the ratios between the diffracted intensities with an incident beam with spin up and spin down neutrons. In this way, one has access to the spin density, although measurements are limited to those magnetic reflections for which the magnetic scattering is sufficiently large (so that the precision of the flipping ration measurement is higher).

3.4.2 Multipolar Spin Density Model

The easiest situation is when the orbital contribution is zero and the magnetization density depends only on the spin. Like the atomic expansion of the electron charge density of Eq. (3.7), one can partition the magnetization spin density $\mathbf{m}_s(\mathbf{r})$ in terms of pseudoatomic contributions:

$$\mathbf{m}_s(\mathbf{r}) = \sum_j \langle \mathbf{s}_j \rangle \langle \sigma_j(\mathbf{r} - \mathbf{r}_j) \rangle \tag{3.11}$$

where $\langle s_j \rangle$ is the spin state of atom j and $\langle \sigma_j(r - r_j) \rangle$ is the thermally averaged spin density of the same site. In analogy with the electron charge density, one may refine a spin-density spherical atom model, where only a monopole spin density is considered, or otherwise calculate a more sophisticated model, like the multipolar model.

The first examples of spin density expanded in terms of atomic multipoles, were provided by Brown et al. (1979) and Boucherle et al. (1982).

In the recent years, the possibility to combine electron charge and electron spin density has emerged. This follows a roadmap designed by Becker and Coppens (1985) and put into practice by Deutsch et al. (2012, 2014), who reported on the first simultaneous refinement of the charge and the spin density in position space, by refining a multipolar model against structure factors from X-ray, unpolarized and polarized neutron diffraction. The use of Bragg X-ray diffraction intensities is pretty obvious for the construction of a charge density model while the non-polarized neutron diffraction data enable fixing with more precision the atomic displacement parameters and the positional parameters of all atoms (especially H atoms). This reduces the inherent correlation between multipoles, fractional coordinates and atomic displacement parameters when using only X-ray diffraction. The flipping ratio from polarized neutron diffraction, enabled refining the spin component of the electron density distribution, so that:

$$\rho_{X-ray}(\mathbf{r}) = \rho^e(\mathbf{r}) = \rho^\alpha(\mathbf{r}) + \rho^\beta(\mathbf{r}) \tag{3.12a}$$

$$\rho_{PND}(\mathbf{r}) = \sigma(\mathbf{r}) = \rho^\alpha(\mathbf{r}) - \rho^\beta(\mathbf{r}) \tag{3.12b}$$

The Hansen and Coppens (1978) formalism was modified (Deutsch et al. 2012):

$$\begin{aligned}
\rho_i(\mathbf{r}) = {} & \rho_{i,core}(\mathbf{r}) + P^\alpha_{i,valence}\left(\kappa^\alpha_i\right)^3 \rho_{i,valence}\left(\kappa^\alpha_i(\mathbf{r} - \mathbf{r}_{0,i})\right) \\
& + \sum_{l=0}^{l_{max}} \left(\kappa'^\alpha_i\right)^3 R_{i,l}\left(\kappa'^\alpha_i(\mathbf{r} - \mathbf{r}_{0,i})\right) \sum_{m=0,l} P^\alpha_{i,lm\pm} y_{lm\pm} \\
& + P^\beta_{i,valence}\left(\kappa^\beta_i\right)^3 \rho_{i,valence}\left(\kappa^\beta_i(\mathbf{r} - \mathbf{r}_{0,i})\right) \\
& + \sum_{l=0}^{l_{max}} \left(\kappa'^\beta_i\right)^3 R_{i,l}\left(\kappa'^\beta_i(\mathbf{r} - \mathbf{r}_{0,i})\right) \sum_{m=0,l} P^\beta_{i,lm\pm} y_{lm\pm}
\end{aligned} \tag{3.13}$$

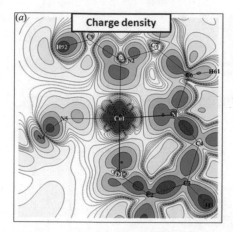 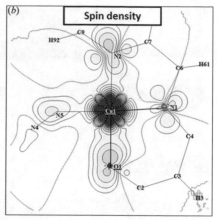

Fig. 3.13 Left: the deformation charge density of $Cu_2L_2(N_3)_2$ [L = 1,1,1-trifluoro-7-((dimethylamino)-4-methyl-5-aza-3-hepten-2-onato)]. **Right**: the electron spin density distribution in the same plane. Reproduced with permission from Deutsch et al. (2014) under Creative Commons Attribution License (https://doi.org/10.1107/S2052252514007283)

Deutsch et al. (2014) successfully refined this model for a metal complex dimer featuring spin active Cu^{2+} ions bridged by two azido ligands (a compound first reported by Aronica et al. 2007), see Fig. 3.13. Some precautions are necessary to avoid divergence and reduce the correlation among parameters, given that the radial part of the density functions may be poorly sensitive to the spin component. Therefore, the restrictions $\kappa_i^\alpha = \kappa_i^\beta$ and $\kappa'^\alpha_i = \kappa'^\beta_i$ are sensible. Of course, the unit cell electroneutrality constraint must hold true anyway and the values of the effective atomic spins must add up to the macroscopic magnetic moment of the sample. An additional constrain may also be applied to the total spin density, which should integrate to the total number of unpaired electrons for the expected electronic spin state of the molecule.

The extension of this model and experimental protocol for a simultaneous refinement of charge and spin electron densities (see Fig. 3.14), both in position and momentum space, using also Compton scattering and magnetic Compton scattering, is the goal of the research in this field.

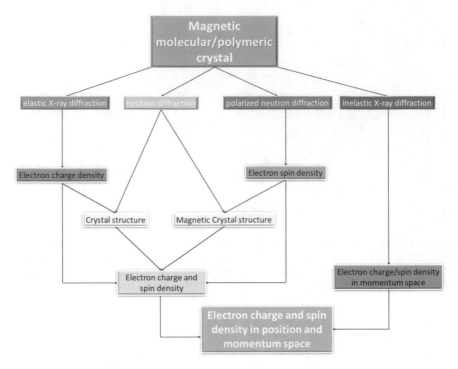

Fig. 3.14 A schematic view of the interplay between different techniques to obtain crystal structures, magnetic structures, electron charge and spin density distributions in position or in momentum space. Noteworthy, in order to obtain electron spin density in momentum space, the inelastic magnetic scattering is necessary

3.5   Transferable Electron Density Models

The model proposed by Hansen and Coppens (1978) was designed with the purpose of comparing pseudoatom multipolar expansions in similar chemical environments. In fact, a *local coordinate system* is adopted for each pseudoatom (see Fig. 3.15),

Fig. 3.15 The local coordinate systems defined for atoms of the molecule of serine in its crystal (zwitterionic) form. One axis (normally z) is typically aligned with the main bond formed by the atom

instead of using the unit cell coordinate system of the crystal (thus the same for all atoms), as typically adopted for the atomic displacement parameters and as used by Stewart (1976) for his multipolar model.

The definition of a local coordinate system finds two possible applications. On the one hand, the comparison of electron density descriptors for chemically equivalent atoms provides a reference frame to better understand how perturbations (like different crystal field effects) affect the atomic properties. This is a natural extension of a simple geometrical analysis of the structure. On the other hand, the average among several pseudoatom expansions offers a robust estimation of the electron density expected around a given atom type. This is useful for refining structures using an aspherical model much more accurate than IAM, even in the absence of the data accuracy necessary for an ad-hoc multipolar refinement.

The mainstream of the improved modelling has originated a very lively field of research. The name which best represents the procedure is the *transferable aspherical atomic model* (TAAM), born after an intuition of Brock, Dunitz and Hirshfeld (1991), who recognized the similarities among same fragments belonging to different molecules and embedded in different crystals. The idea further developed toward the construction of databanks of pseudoatoms (the first one was proposed by Pichon-Pesme et al.1995). Each of them statistically represents the whole population of atoms belonging to a specific type. The pseudoatom parameters enable constructing an aspherical atomic form factor to be used for a structural refinement instead of the standard spherical form factor, thus improving the quality of the model, without extra expenses for the data collection. A side benefit is a good approximation of the electrostatic properties of the molecule under investigation and of the electrostatic interactions originated by the specific crystal packing.

Several schemes have been proposed for TAAM, as summarized in Table 3.2. There are two main pathways to construct the databanks, by averaging: (a) parameters refined on model systems measured experimentally; (b) parameters refined from theoretically computed molecular electron densities. The definition of atom types to be included as separate entries in the databank is mainly driven by chemical intuition, without a full statistical analysis. A conceptually different scheme is using pseudoatom aspherical form factors calculated by defining a single reference molecule for each atom, preserving the local stereochemistry but reducing the size of atom substituent groups. This scheme is called invariant atom model (*invariom*). Of course, the threshold to cut the length of a group bound to the target atom is arbitrary and many possible recipes could be adopted, although the standard rule to saturate the first neighbor atoms with H atoms was proved to be sufficiently accurate, thus reducing the complexity of the databank. The invariom method has been recently included in the most popular structure refinement program (SHELX), see Lübben et al. (2019).

The transferability of form factors implies some assumptions. The monopole populations of the *pseudoatoms* must be normalized for electroneutrality of the crystal unit cell perhaps with some specific constraint on each molecule in the unit cell, in case of co-crystals, or each ion, in case of salts. Another issue is the proper

Table 3.2 A summary of the main TAAM proposed in the literature.

Authors	Databank name	Databank type
Pichon-Pesme et al. (1995)	experimental library multipolar atom model (**ELMAM**)	Experimental multipolar parameters, averaged from a series of experimentally refined multipolar models on representative molecules
Domagała et al. (2012)	**ELMAM2** (an extension)	Same as **ELMAM**
Volkov et al. (2004)	University of Buffalo databank (**UBDB**)	Multipolar parameters refined from theoretical calculations on a series of representative molecules
Jarzembska and Dominiak (2012)	**UBDB2011** (an extension)	Same as **UBDB**
Kumar et al. (2018)	**UBDB2018** (an extension)	Same as **UBDB**
Dittrich et al. (2004)	Invariant atom (**INVARIOMS**)	Multipolar parameters refined from theoretical calculations on uniquely defined molecules for each atom type, based on a topological connectivity

orientation of the transferable pseudoatom. In fact, while in the databank the pseudoatom expansion is constructed assuming an ideally symmetric stereochemistry, the atom in the crystal structure may experience a distorted steoreochemical environment. This means that the databank pseudoatom coordinate system only approximately fit the local coordinate system. Practically some assumption with respect to the orientation of the main axis (and if needed of the second main axis) must be adopted. Only recently, an unbiased scheme, based on the coincidence of the local and database inertial tensor has been proposed, although applied for the transferability of atomic polarizabilities, instead of multipoles (Ernst et al. 2019).

The TAAM found applications for accurate modelling of (mainly) biomolecules, for example the refinement of protein structures (Jelsch et al. 2000, Guillot et al. 2008), which is a very demanding and extraordinary task. The software MOPRO (Jelsch et al. 2005) has been extended and adapted to handle this kind of structural models, including the treatment of disorder.

More recently, the TAAM approach was adopted also for the possible refinement of crystal structures form electron diffraction experiments (Gruza et al. 2020).

Chapter 4
Testing Chemical Bond Theories with Quantum Crystallography

Can we see the chemical bonds?

The chemical bond is not a quantum mechanical observable. The sticks linking spheres (atoms) in experimental crystal structures are automatically drawn by a visualization software when the two atomic nuclei are close enough, depending on some predefined criterion, but they do not imply any physical observation of the chemical bonds. However, theories based on the scalar field of functions like the electron density allow the visualization and the quantification of entities (sometime, genuine observables) that are clues of chemical bonding.

In other words, we can somehow observe those sticks linking atoms and we can even infer the nature of the bonds and therefore correlate quantum mechanical observables with purely theoretical speculations on the chemical bond, like octet rules leading to the Lewis structures (Lewis 1916).

4.1 Quantum Chemical Topology

Under the name of Quantum Chemical Topology (QCT) (coined by Popelier and Aicken 2003) fall several theoretical approaches based on a topological treatment of scalar functions, like the electron density, the electric potential, the electron localization function, etc. The purpose of these treatments is quantifying features of the scalar function to enable a mathematical description. More precisely, Popelier (2016) defined QCT as

> a branch of theoretical chemistry that uses the language of dynamical systems (e.g. attractor, basin, gradient path, critical point and separatrix) to partition chemical systems and characterize them via associated quantitative properties, e.g., atomic charge or bond order.

Topology is not a novelty in chemistry, nor in crystallography. The Lewis structures themselves are simple examples of chemical topology, where the graph

© The Author(s), under exclusive license to Springer Nature Switzerland AG 2022
P. Macchi, *Quantum Crystallography: Expectations vs Reality*,
SpringerBriefs in Crystallography, https://doi.org/10.1007/978-3-030-95641-7_4

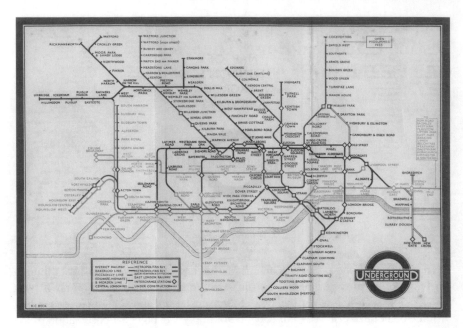

Fig. 4.1 The map of the London underground, drawn by Harry Beck in 1933 (reproduced with permission of the London Transports Museum)

addresses the connections among atoms and visualize rules based on a well-established theoretical framework. In crystallography, many topological approaches are in use, for example to classify supramolecular interactions like the hydrogen bond graph (Grell et al. 2002) or the inorganic nets (Wells 1977).

The topology plays the role of a reductionist thinking, with the purpose of simplifying the description of the geometry of an object making it more understandable and comparable with others. A huge array of data is rationalized (i.e. simplified and classified) based on a mathematical tool. See Shevchenko and Blatov (2021) for applications to crystallography.

A seamless example of the simplification obtainable with a topological graph is the underground map, like the famous London underground map drawn in 1931 by Harry Beck (see Fig. 4.1). He introduced three novelties in the map: (a) surface details were removed (the only exception for the London tube was the location of the river Thames); (b) the railway paths are idealized with straight lines and smooth turns of 45° or 90°; (c) all stations are equally distant from the next ones.[1]

Obviously, a topological simplification must be taken *cum grano salis*. One cannot use the topological map as a real map of the city because the surface paths may differ substantially. Translated into the realm of quantum crystallography, the topology has the function of simplifying the information, which is otherwise too difficult to

---

[1] Inspiration for this example came after attending a lecture by A. Baricco (*The London Underground Map – About the truth*) at the Mantua Literature Festival (2016).

extract from a set of multivariate data, like a full description of the electron density distribution or a full list of atomic coordinates. This does not concern QCT only, but it holds true for other applications of topology in crystallography. For example, the topological analysis of nets (Wells 1977) is undoubtedly an elegant way to describe the structure of polymeric compounds, simplifying the structural complexity and allowing easier comparisons among crystals of very different nature and type: the graph is made of lines and nodes that do not necessarily represent the bonding network of the crystal structure; the monomeric units may be reduced to just one node, despite consisting of many atoms; all bonds are equal, although being of very different nature. Pros and cons are obvious: we have a rationalization of the structure, but we cannot predict the properties of the materials; we have a powerful mean to recognize similarities and differences among structures, but we cannot extrapolate at which other level those similarities remain valid. For example, two completely different polymeric structures may belong to the very same topological class, like two completely different compounds may crystallize in the same space group type. The properties of the two compounds may however be extremely different.

For many topological approaches, the simplification enables a better understanding, but it does not provide predictive power. For this reason, it is often argued that topology in chemistry is *descriptive* but not *predictive*. Danovich et al. (2013) remarked this especially for the QCT methods. Is this criticism correct?

4.1.1 Quantum Theory of Atoms in Molecules and in Crystals

In QCT, the Morse theory (Morse 1925) is applied like in other topological approaches, but the hypothesis underneath is that there is a quantum mechanical ground for the topological objects (attractor, basin, gradient path, critical point and separatrix).

Within QCT, the most famous theory, and the most adopted in experimental studies, is that proposed by Bader (1990), namely the Quantum Theory of Atoms in Molecules (QTAIM), which becomes the Quantum Theory of Atoms in Molecules and Crystals (QTAIMAC) for periodic systems (Gatti 2005). The main role is played by the scalar function electron density $\rho(\mathbf{r})$ and its gradient vector field $\nabla\rho(\mathbf{r})$, used to partition the space into (atomic) *domains*. They do not overlap and are separated by (interatomic) surfaces, whose points are connected by special $\nabla\rho(\mathbf{r})$ trajectories, that are perpendicular to the surface normal, originate at infinite, and terminate at saddle points of the electron density. These are the so-called *bond critical points* (*bcp*). Instead, the gradient trajectories passing through all other points terminate at electron density maxima, called *attractors*. Normally an attractor coincides with a nuclear position, where in fact the electron density features a cusp. However, additional attractors may occur far from nuclear positions (*non-nuclear attractors*, i.e. electron density maxima at sites which do not coincide with any atomic nucleus). At the bond critical points, as well as at the attractors, the gradient of the electron density vanishes. The formers are characterized by two negative and one positive curvatures, whereas the latter have three negative curvatures. The two gradient trajectories originating from a *bcp* terminate at each of the two attractors forming the *bond path*, a line of maximum electron density linking the two attractors (like the ridge of a mountain connecting two peaks).

Topological rules require the presence of other saddle points, characterized by one negative and two positive curvatures, and minima, characterized by three positive curvatures. The formers are necessarily found in the presence of a loop, and therefore called *ring critical points* (*rcp*). The minima are instead necessarily associated with four or more loops sharing edges and are called *cage critical points* (*ccp*). For a finite system, the Poicaré-Hopf relationship holds (Hopf 1916):

$$N_{max} - N_{bcp} + N_{rcp} - N_{min} = 1 \qquad (4.1)$$

N_{max}, N_{bcp}, N_{rcp}, and N_{min} are the numbers of electron density maxima, bond critical points, ring critical points, and minima. For an infinite periodic system, Eq. (4.1) becomes (Jones and March 1985):

$$N_{max} - N_{bcp} + N_{rcp} - N_{min} = 0 \qquad (4.2a)$$

with

$$N_{max} \geq 1; N_{bcp} \geq 3; N_{rcp} \geq 3; N_{min} \geq 1 \qquad (4.2b)$$

where the number of points refers to the points found in a unit cell. Moreover, some special symmetry sites in a crystal imply the presence of critical points. For example, an inversion center is necessarily a critical point because $\nabla \rho(\mathbf{r}) = 0$.

So far, we briefly discussed the mathematical foundations, but where is the quantum mechanical ground? One of the arguments stressed by Bader is the possibility to recognize atoms as *proper open subsystems* of a molecule or a crystal. According with Bader (1990):

> The notion that a molecule can be viewed as a collection of atoms linked by a network of bonds, a notion that has already been shown to be rooted in the topological properties of the charge distribution, is the operational principle underlying our classification and understanding of chemical behaviour. It would appear, therefore, that to find chemistry within the framework of quantum mechanics one must determine the values of observables for pieces of a total system, i.e. for a subsystem.

In principle, any partitioning of the space is possible, but Bader stressed that only one based on the interatomic surfaces (that guarantee a zero flux of the density through them) is quantum mechanically grounded. Thus, a quantum mechanical atom, or better a quantum mechanical *atomic basin*, is the volume enclosed by zero flux surfaces and containing one, and only one, attractor.

Some criticism was raised, for example by Cassam-Chenäi and Jayatilaka (2001), because of the non-uniqueness of the definition, which was rebounded by Bader

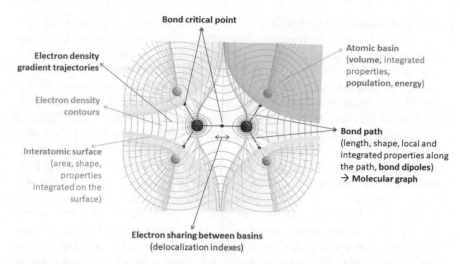

Fig. 4.2 Schematic representation of the information available from a topological analysis of the electron density

(2001) who stressed that "an atom is defined as the union of an attractor and its basin". Bader stressed that the conditions of the QTAIM theory complies with Schwinger's action principle (Schwinger 1951a, 1951b; valid also in case of proper subsystems like the open QTAIM atom). For non-nuclear maxima, the combination of the attractor and the corresponding basin gives rise to the *pseudo-atoms* (Bader 1990).[2]

The atom in molecule concept could be transferred to periodic systems as well (Gatti et al. 1994), though with an important difference. Atoms in molecules have, in general, infinite volumes because the trajectories contained in the atomic basins and attracted by the local maximum originate at infinite (exceptions are atoms encapsulated in cages). In ideal crystals, instead, the atomic volumes are necessarily finite, whereas in real crystals atoms on the surfaces have infinite volumes. In crystals, it occurs more frequently to find non-nuclear attractors, especially in solids consisting of atoms with very diffuse valence shells (like metals, alloys, etc.). One seminal example is the metal Be (Iversen et al. 1995).

In Fig. 4.2, the main quantities of QTAIM and QTAIMAC are represented schematically on the molecule of C_2H_4. Apart from the topological features, QTAIM returns vital information on how chemical bonding establishes among atoms. In

---

[2] It should be noted that the term *pseudo-atom* is also widely adopted to indicate the set of nuclear-centered multipole functions used in the multipolar model. The name comes from Stewart (1976) and it should not be confused with that by Bader. Interestingly, in *electrides*, i.e. solids in which electrons occupy interstitial regions in a crystal and behave as anions, the electron density is a local maximum, not associated with a nuclear site. Without invoking a rigorous QTAIM or QTAIMAC definition, Miao and Hoffmann (2014) named these special entities *quasi-atoms*, with a definition that seamlessly coincide with Bader's *pseudo-atoms*.

particular, the theory has been widely adopted as the paradigm for classifying chemical bonds and for the interpretation of their nature. While the electron density partitioning provides a quantum mechanical definition of an atom, many other quantities can be mapped in this subspace and provide precious indications on the bonding state of an atom. The electron density itself, if integrated in the atomic basin, returns the number of electrons belonging to that atom, hence the atomic charge. The integration of the pair-density, instead, enables the evaluation of electron localization and delocalization indices. The pair density, or two-electron density $\pi(\mathbf{r}_1, \mathbf{r}_2)$, is an extension of the one-electron density, i.e. $\rho(\mathbf{r})$ itself, the quantity so far discussed. While the $\rho(\mathbf{r})$ is the probability to find one electron at the position $\mathbf{r}$, $\pi(\mathbf{r}_1, \mathbf{r}_2)$ is the probability to find (any) one electron at $\mathbf{r}_1$ and any other electron at $\mathbf{r}_2$. Deviations of $\pi(\mathbf{r}_1, \mathbf{r}_2)$ from a simple stochastic expectation, $\pi(\mathbf{r}_1, \mathbf{r}_2) \propto \rho(\mathbf{r}_1)\rho(\mathbf{r}_2)$, indicate the occurrence of a correlation between electrons (the pairing). Counting the electron pairs which are shared between two (or more) atomic basins is an indication of chemical bonding. Blanco et al. (2005) pushed further this concept including a partition of energies in terms of atomic contributions, in the so-called *interacting quantum atom* approach, which also distinguishes the nature of the interaction between two atoms as due to classical electrostatics and non-classical quantum mechanical exchange.

Many chemical bonding analyses also make use of another kind of density, which is the density of electronic energy, a position-dependent scalar function that could be used like the electron density itself. Bader (1990) has shown that the distribution of potential energy density mimics the electron density distribution because a bond path in the electron density is mirrored by an equivalent quantity in the potential energy density distribution. Cremer and Kraka (1984) especially stressed that the total energy density distribution (potential + kinetic) may be an indicator of the nature of the chemical bonding, because a negative value (especially at the bond critical point) implies an excess of potential energy, a fingerprint of covalency. Although this criterion has been widely adopted in analysis of intra-molecular and intermolecular chemical bonds (see for example Macchi and Sironi 2003) the concept has rarely been adopted to identify molecules in crystals (Tsirelson 2002; Krawczuk and Macchi 2014). An example of total electron density map in the molecule of C_2H_4 is shown in Fig. 4.3.

While these approaches are undoubtedly powerful and informative, they can hardly find evidence from experiments, which are not able to directly reveal the pair density or the energy densities if the model adopted is the multipolar model. Approximations are possible (see Tsirelson 2002; Fig. 4.4), but they have a restricted range of validity and will not be further discussed here. As we will see in Chap. 5, new methodologies enable refinement of some coefficients of the electron density matrix (a descriptor of the quantum state of a system) or the calculation of an experimentally restrained wavefunction. In both cases, information on the electron delocalization between atoms or the exact energy density distributions become available. Calculations of X-ray restrained wavefunction or refinement of electron density matrices are relatively recent and more complicated than a multipolar model. Therefore, they are not broadly employed to analyze experimental measurements. Moreover, the significant engagement of theory gives the (incorrect) impression that wavefunction or

Fig. 4.3 Top: The Laplacian of the electron density distribution, $\nabla^2\rho(\mathbf{r})$, of ethylene (negative regions are bold contours; positive regions are dotted contours). **Bottom**: The total electron energy density (negative regions are bold contours; positive regions are dotted contours)

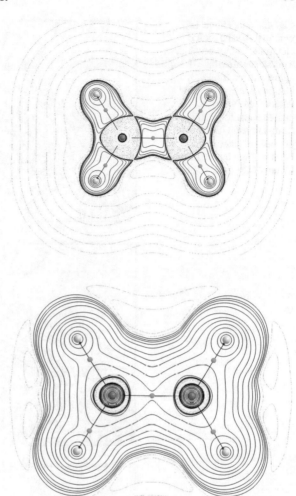

density matrix approaches are less experimental than a multipolar model. In facts, in both cases a significant amount of theory is adopted.

For this reason, many researchers tried instead to extract information on the electron sharing directly from one-electron properties, directly available from the electron density in a multipolar expanded form. In particular, the analysis of the second derivatives of the electron density (the Laplacian, $\nabla^2\rho(\mathbf{r})$) or of the electric potential (and its derivatives) have been quite deeply exploited in order to solve the puzzle of the chemical bond and reveal its nature.

The analysis of the Laplacian of the electron density (see Fig. 4.3) is very common, because it inherently reveals regions of electron density accumulation ($\nabla^2\rho(\mathbf{r}) < 0$), typical for strong covalent bonds. The simple inspection of $\nabla^2\rho(\mathbf{r})$ at the bond critical

Fig. 4.4 The kinetic (**a**), potential (**b**) and total **c** electronic energy densities in a crystal of urea, approximated from the electron density functions determined from experiments. Reproduced from Tsirelson (2002) with permission guaranteed by the license of the International Union of Crystallography

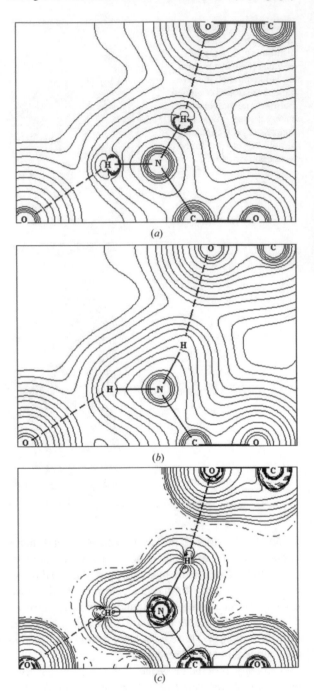

point, though, may be deceptive and could lead to inconsistent conclusions. For a more in-depth discussion, see Macchi and Sironi (2003).

The abuse of indicators derived from QTAIM and QTAIMC analyses has been criticized, for example by Dittrich (2017) who questioned the possible future of topological analysis of experimental charge density. Dittrich focused on the production of too many tables and table entries, reporting results of topological analyses. It often occurs that bonding descriptors are listed also for chemical bonds that are of poor interest in the economy of a study on a given molecule/crystal. This produces a huge level of noise, making it difficult to extract the significant information obtained from a QCT analysis and appreciate the real innovation grade of the research.

Although addressing a serious problem and a potentially serious pitfall, Dittrich did not consider that the whole modern research is similarly affected by an increasing noise level of mere incremental science. If carried out with intelligence, topological analysis is very informative and useful (see more in details Macchi 2017).

4.1.2 Other Scalar Fields

Electron density is not the only scalar field used in QCT. Quantities directly derived from $\rho(\mathbf{r})$ are also subject of similar investigations. For example, the Laplacian of the electron density, $\nabla^2 \rho(\mathbf{r})$, can be analyzed in terms of critical points, of course with a completely different meaning. Minima of $\nabla^2 \rho(\mathbf{r})$ represent the largest local concentration of charge density, that one can associate with bonding or non-bonding electron pairs. The set of critical points around an atom defines the so-called *atomic graph*, including charge density depletions (local maxima of $\nabla^2 \rho(\mathbf{r})$) and saddle points.

An analysis very similar to the QTAIM analysis of $\rho(\mathbf{r})$ is that of the electric potential $V(\mathbf{r})$, classically derived from the electron density:

$$V(\mathbf{r}) = \int \frac{\rho^{tot}(\mathbf{r}')}{|\mathbf{r} - \mathbf{r}'|} d\mathbf{r}' = \sum_i \frac{Z_i}{\mathbf{r} - \mathbf{R}_i} - \int \frac{\rho^{electron}(\mathbf{r}')}{|\mathbf{r} - \mathbf{r}'|} d\mathbf{r}' \qquad (4.3)$$

where Z_i and $\mathbf{R}_i$ are respectively the nuclear charge and the nuclear position of atom i. $\rho^{tot}(\mathbf{r}')$ is therefore the *total charge density*, a quantity that cannot be analyzed directly (given the discrete distribution of the nuclear charges in contrast with the continuous distribution of the electron charge), but indirectly through the electrostatic potential.

The algorithms to calculate of electrostatic maps from X-ray diffraction and a multipolar expansion are exhaustively explained by Stewart (1979, 1982), Su and Coppens (1992), Ghermani et al. (1993), Bouhmaida et al. (1997) and will not further discussed here.

The gradient of $V(\mathbf{r})$ is the electric field, that can be used like the electron density gradient to partition the space into atomic domains. The most important difference, and very peculiar, is that atomic domains calculated using $V(\mathbf{r})$ are necessarily electrically neutral. This is a consequence of Gauss divergence theorem. Quantum crystallographic applications, based on the $V(\mathbf{r})$ partition, are less exploited (one of the few examples was published by Bouhmaida et al. 2002).

Scalar fields based on the electron and the energy densities are accessible exactly only through the wavefunction. With a multipolar model, they can only be approximated (see Tsirelson 2002). Among these fields, one can mention the very famous electron localization function (ELF by Becker and Edgecombe 1990):

$$ELF(\mathbf{r}) = \frac{1}{1 + \chi^2(\mathbf{r})} = \frac{1}{1 + \left(\frac{D_\sigma(\mathbf{r})}{D_\sigma^0(\mathbf{r})}\right)^2} \qquad (4.4a)$$

where $D_\sigma(\mathbf{r})$ is the difference between the kinetic energy density $G_\sigma(\mathbf{r})$ (positively definite) and the kinetic energy density of a Boson (von Weizsäcker 1935), where σ indicates one spin type. Therefore, $D_\sigma(r)$ can be interpreted as the excess of kinetic energy of Fermions. The reference $D_\sigma^0(r)$ is taken as the corresponding value for a uniform electron gas with spin density equal to $\rho(\mathbf{r})$. By definition, $0 \leq ELF \leq 1$ and it follows that an ELF value of 1.0 indicates a perfect electron localization, whereas 0.5 corresponds to a perfectly homogeneous gas. In a more general formulation proposed by Savin et al. (1992), a total density replaces the density of a given spin type. The topological analysis of ELF is quite different from that of the electron density, given that the chemical bonding itself produces a domain and the value of the localization function can be used in fact to characterize the nature of the bond.

In the recent years, other functions have been proposed that can, like the ELF, provide information on the electron localization, but for sake of brevity they cannot be considered here.

4.2 Applications of Quantum Chemical Topology

All the methods discussed above are often used in quantum crystallography studies to investigate a system and analyze, qualitatively and quantitatively, the interactions that stabilize atoms in a molecule and molecules in a crystal.

QTAIMAC approaches were introduced in the early 1990s thanks to the popularization of QTAIM theory and the diffusion of the multipolar model. Many studies appeared and contributed to validate theoretical simulations and extend the range of systems investigated wit QTAIMAC, providing more substantial evidence for the translation of QTAIMAC concepts in chemical language, an operation which was not so easy and straightforward (see for example the discussion in Macchi and Sironi 2003).

The QTAIMAC analysis of the scalar electric potential was also proposed (Bouhmaida et al. 2002), although it received less attention, and it is not often adopted.

For other theories based on scalar fields, instead, the application to experimentally derived functions was not so straightforward. As mentioned above, ELF requires calculation of the kinetic energy density, which is quite complex because it requires the knowledge of the wavefunction and not only of the electron density. However, some successful approximations were proposed to calculate the kinetic energy density from the electron density and its derivatives (Abramov 1997). This paved the way towards more studies on experimentally derived ELF functions, as well as on the analysis itself of the kinetic or the total electronic energy density.

4.2.1 Statistical Use of Quantum Crystallographic Information

In the crystallographic community, a popular method is the *structure correlation analysis*. The progressive accumulation of data on molecular structures, obtained from experimental crystal structure determinations, inspired Bürgi and Dunitz (1983) who proposed to correlate various geometries of a molecule or a molecular fragment, as determined in different crystals in order to obtain a qualitative picture of the potential energy surface of the molecule/fragment. In other words, the responses of a molecule/fragment is sampled in as many as possible crystal structures, differing of course in the extent and direction of the crystal field acting on the molecule/fragment as well as on the chemical environment of the fragment. This approach enabled to draw interesting conclusions for example on the mechanism of some chemical reaction, being more reliable and rapid than using theoretical methods, at that time, computationally expensive. The hypothesis underneath is that every intermolecular contact occurring in a crystal is in principle an incipient chemical reaction, frozen by the stabilization induced through the crystallization, but variable as the chemical environment changes.

The structure correlation method is still widely adopted, thanks to the growth of the databases containing organic/organometallic crystal structures. This method is inherently quantum crystallographic, because from crystallographic information one can retrieve quantities such as the molecular or intermolecular potential energy surfaces, albeit approximate and empirical. Even more important is that the same approach could be valid also for correlating the electron density changes in response to the modified surrounding of a molecule, thus going well beyond the simple geometrical features. For example, Macchi et al. (2002) studied the *metamorphosis* of carbonyl ligands in transition metal clusters, when changing the binding mode from terminal to bridging modes of coordination. Following the same principles of the structure correlation method, the evolution can be followed experimentally by measuring the electron density on various crystals chosen along the interconversion path emerging from the database search (see Fig. 4.5). The added value of analyzing the electron

Fig. 4.5 The structure correlation analysis in combination with multiple QTAIMC analysis of the fluxional behavior of carbonyl ligands in transition metal dimers. Reproduced from Macchi and Sironi (2003) with permission of Elsevier

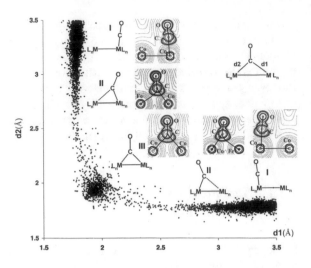

density changes is that one can easily realize how a localized donor–acceptor interaction (typical of the terminal carbonyl) continuously transforms into a delocalized 3-center-4-electron bonded system in the bridging mode, going through a semi-bridging mode. Despite an abrupt change of the molecular graph, there is no discontinuity in many other parameters. This reinforces the conclusion that the topology of the electron density does not reveal alone the grounds of chemical bonding, because a bond path cannot tell us how many electrons are in fact shared among the bonded atoms or if *through bond* delocalized mechanism accompanies the *through space* direct interaction. Noteworthy, this information is seamlessly conveyed by the electron delocalization indices, as discussed above, but their experimental derivation is not possible using the multipolar model.

Another example of quantum crystallographic structure correlation analysis was previously shown by Espinosa et al. (1998), who correlated several experimental electron densities determined for a variety of hydrogen bonds (HB) in crystals (see Fig. 4.6). Again, depending on the crystal structure and on the modification of the chemical bonds to the HB donors or acceptors, the electron density, and its derivatives, respond in a rather continuous way. This is of course in keeping with the idea that the hydrogen bond is nothing else than an interrupted protonation reaction and a structure correlation reveals the reaction path of such protonation. The study by Espinosa et al. (1998) enabled an interesting, albeit empirical, estimation of the stabilization energy provided by the HB, which depends on the local potential energy density evaluated at the bond critical point located in between the donor and the acceptor.

More recently, Hupf et al. (2017) used a similar approach to map the trajectories of nucleophilic substitution in inorganic complexes, combining the geometrical, spectroscopic and QTAIM indices along a pseudo-reaction path.

Fig. 4.6 The correlation between kinetic and potential energy densities calculated at bond critical points of intermolecular hydrogen bonds, from experimental multipolar models. Reproduced from Espinosa et al. (1998) with permission of Elsevier

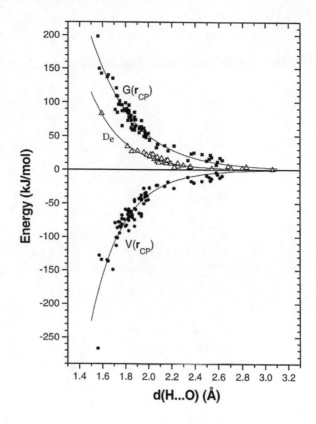

Chapter 5
Calculating the Crystal Wavefunction from Crystallographic Experiments

Can we measure the wavefunction?

In quantum mechanics, the wavefunction is a fundamental quantity on which the entire theoretical framework is grounded. Together with the corresponding operators, the wavefunction enables the calculation of expectation values of the observables. This addresses the two conceptual problems of quantum mechanics: the physical meaning of the wave function and of the measurement.

In a strict interpretation of quantum mechanics postulates, the wavefunction is not an observable. A classical interpretation refers to the so-called *collapse* of the wavefunction when a measurement is performed. If the position is measured the wavefunction localizes at the corresponding coordinate in the position space, returning the probability to find the electron at that point. However, this affects any subsequent measurement of the momentum, because the wavefunction has been perturbed and the measured coordinate in momentum space would not correspond to the previous measurement.

Therefore, the most straightforward answer to the question whether we can observe a wavefunction would be a simple and strict: "no". Nonetheless, the *quantum state tomography* is a known method proposed to measure *indirectly* the wavefunction by probing it with known quantum states (Vogel and Risken 1989). More recently, Lundeen et al. (2011) proposed also a *direct* method based on two subsequent measurements: a *weak* measurement which does not perturb the wavefunction followed by a *strong* measurement. Thus, the orthodox perspective of the wavefunction being non-measurable is challenged by methods to circumvent the problem, which finds enormous interest in the exploding field of quantum information and quantum computation (Nielsen and Chuang 2010).

© The Author(s), under exclusive license to Springer Nature Switzerland AG 2022
P. Macchi, *Quantum Crystallography: Expectations vs Reality*,
SpringerBriefs in Crystallography, https://doi.org/10.1007/978-3-030-95641-7_5

5.1 Early Attempts

The above-mentioned methods are mainly connected with quantum optics, while little is known outside the crystallography community about methods that make use of Bragg diffraction from crystals. Instead, a historical background needs to be shared among the scientific community and better emphasized. It is worthwhile to start from a statement by one of the fathers of modern chemistry, Linus Pauling (1926), writing a letter to the head of his department:

> I think that it is very interesting that one can see the ψ-functions of Schrödinger's wave mechanics by means of the X-ray study of crystals. This work should be continued experimentally; I believe that much information regarding the nature of the chemical bond will result from it, and I am hoping that it might be possible to do the experimental work in Gates.

Pauling was of course attracted by the possibility to learn more on chemical bonding and was driven by the early studies by Bragg et al. (1922), see Chap. 2. Pauling's statement was premonitory and visionary, going beyond the (limited) possibilities available in 1926. It took three decades to see a partial realization of this vision and more than half a century to put into practice.

Since the early days of electron density analysis, an important goal was the direct refinement of wavefunction coefficients from X-ray diffraction data. This goal dates back to the work of Weiss and De Marco (1958), who attempted to determine the electronic configurations of metal atoms in their elemental solid forms, a task complicated by the difficult refinement of electronic populations in the outermost shells, in keeping with the discussion in Chap. 3 (see Fig. 3.1). Weiss and De Marco did not attempt to calculate a wavefunction from X-ray diffraction, however they tried to assign electron population coefficients to the atomic orbitals of the atoms in the crystal. Some years later, Weiss (1966) went well beyond when he proposed to use the diffracted intensities to correct the theoretically computed wavefunctions using a kind of perturbative approach. The true Hamiltonian could be regarded as the Hartree-Fock Hamiltonian corrected by a term based on the X-ray diffraction measurements:

$$\mathcal{H} = \mathcal{H}^0 + \lambda\mathcal{H}' = \mathcal{H}^{Hartree-Fock} + \lambda\mathcal{H}^{X-ray\ correction} = \mathcal{H}^{HF} + \lambda\mathcal{H}^{XRD} \quad (5.1)$$

In the perturbative approach, the hypothesis is that the correction $\lambda\mathcal{H}'$ is small compared with $\mathcal{H}^0$, which therefore must be a relatively good zero-order approximation. The Hartree-Fock method (Hartree 1928; Fock 1930) is a kind of mean-field approximation, according to which the function of an electron is calculated in the average field produced by all the other electrons and all the nuclei. This scheme inherently neglects the instantaneous correlation among the electrons, which is instead relevant to determine the electronic energies and the exact electron distribution of a system. However, as discussed in Chap. 2, the Hartree-Fock calculation of atomic form factors is a good approximation (see for example Fig. 2.2). Therefore, Weiss' hypothesis was sensible, and one could use the Hartree-Fock wavefunction as ψ_0^0, i.e. the wavefunction for the unperturbed (^{0}) state ($_0$). In terms of energy, the correction is:

$$\epsilon' = \int \psi_0^0 \mathcal{H}' \psi_0^0 dV \tag{5.2}$$

which is the expectation value of the perturbation mapped over the unperturbed wavefunction of state ($_0$). This holds true not only for the energy, but also for the scattering if, instead of considering the Hamiltonian operator, one considers the scattering operator. This means that using a good theoretical wavefunction for a crystal, one can improve it using the measured diffraction intensities.

The first order correction to the wavefunction of state ($_0$) is:

$$\psi_0 = \psi_0^0 + \lambda \left(\frac{\partial \psi_0}{\partial \lambda} \right)_{\lambda=0} \tag{5.3}$$

where $\left(\frac{\partial \psi_0}{\partial \lambda} \right)_{\lambda=0}$ can be approximated by a summation over all other unperturbed states (which are orthogonal to ψ_0^0), evaluated through the perturbation operator. In case of the Hamiltonian, we obtain the well-known first order perturbation correction to the wavefunction:

$$\left(\frac{\partial \psi_0}{\partial \lambda} \right)_{\lambda=0} = \sum_{i=1}^{n} \frac{\int \psi_0^0 \mathcal{H}' \psi_i^0}{E_0^0 - E_i^0} \tag{5.4}$$

where E_0^0 and E_i^0 are the energies of the unperturbed states 0 and i, respectively. Weiss' suggestion was therefore to employ the scattering operator to evaluate the perturbation of the wavefunction.

Weiss' approach was not translated into a workable algorithm, although contemporarily Mukherji and Karplus (1963) proposed a method to constrain a theoretically calculated wavefunction of a molecule to some experimentally measured observable, like the dipole moment. Although methodologically different from Weiss' proposal, a constrained wavefunction could be similarly regarded as the correction with respect to a theoretical wavefunction, although the corrective term does not necessarily depend on the uncorrected wavefunction. In Fig. 5.1 we see how the two approaches could be schematically summarized. As we will see, Weiss' idea was put into practice only many years later following a scheme that somehow resembles Mukherji and Karplus's one.

Meanwhile, Clinton, Nakhleh and Wunderlich (1969), Clinton, Galli and Massa (1969), Clinton, Henderson and Prestia (1969), Clinton and Lamers (1969) and Clinton et al. (1969) devised a series of equations necessary to construct a pure state density matrix using external constraints and afterwards pioneered the possibility to compute a density matrix from X-ray diffraction intensities (Clinton and Massa 1972; Clinton et al. 1973).

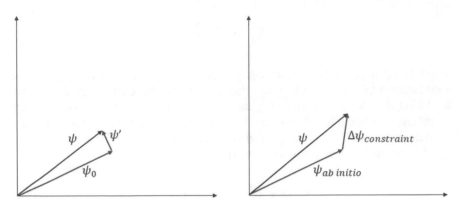

Fig. 5.1 **Left: a** Schematic representation of the perturbative approach. **Right: b** Schematic representation of a constrained procedure

Here we need some more definitions, because the concept of density matrix may not be familiar to a crystallographer. The *density matrix*[1] is a representation of the *density operator*. For a system of N particles, the density operator of a *pure state* has the form of an *outer* product of the state vectors[2]:

$$\hat{\rho} = \begin{pmatrix} \psi_1(\mathbf{r}_1) \\ \dots \\ \psi_n(\mathbf{r}_N) \end{pmatrix} \left(\psi_1^*(\mathbf{r}_1), \dots, \psi_n^*(\mathbf{r}_N) \right) = \begin{pmatrix} \psi_1(\mathbf{r}_1) \\ \dots \\ \psi_n(\mathbf{r}_N) \end{pmatrix} \otimes \begin{pmatrix} \psi_1^*(\mathbf{r}_1) \\ \dots \\ \psi_n^*(\mathbf{r}_N) \end{pmatrix} = \Gamma$$

(5.5)

where $\psi_i^*(\mathbf{r}_i)$ is the complex conjugate form of $\psi_i(\mathbf{r}_i)$. The state is pure because a single vector describes it, having probability of 100%. The pure state density matrix has many important properties. It is idempotent, it is Hermitian and its trace is equal to one:

$$\Gamma^2 = \Gamma$$

(5.6a)

$$\Gamma^\dagger = \Gamma$$

(5.6b)

$$Tr(\Gamma) = 1$$

(5.6c)

The more general case, however, is the system being in a *mixed* state, which means that two or more state functions interplay and each state has a given probability to occur. In this case, the density matrix describes the statistical average of this mixture

---

[1] The name matrix is somewhat unfortunate, because the position and spin coordinates of the particles are continuous variables.

[2] For the sake of clarity, we do not introduce here the most popular bra-ket notation to indicate state vectors and their products.

of quantum states. For N particles, an element of the density matrix Γ^N assumes the form:

$$\Gamma^N(\mathbf{r}_1, \mathbf{r}_2.., \mathbf{r_N}; \mathbf{r'}_1, \mathbf{r'}_2.., \mathbf{r'_N}) = \psi(\mathbf{r}_1, \mathbf{r}_2.., \mathbf{r_N})\psi^*(\mathbf{r'}_1, \mathbf{r'}_2.., \mathbf{r'_N}) \quad (5.7)$$

For the calculation of properties of a system, it is convenient to reduce the density matrix by integrating over possible permutations and consider only the *one-electron* density matrix Γ^1 or the *two-electron* density matrix Γ^2:

$$\Gamma^1(\mathbf{r}_1; \mathbf{r'}_1) = N \int \psi(\mathbf{r}_1, \mathbf{r}_2.., \mathbf{r_N}) \cdot \psi^*(\mathbf{r'}_1, \mathbf{r}_2.., \mathbf{r_N})d\mathbf{r}_2...d\mathbf{r}_N \quad (5.8a)$$

$$\Gamma^2(\mathbf{r}_1, \mathbf{r}_2; \mathbf{r'}_1, \mathbf{r'}_2) = N \int \psi(\mathbf{r}_1, \mathbf{r}_2.., \mathbf{r_N}) \cdot \psi^*(\mathbf{r'}_1, \mathbf{r'}_2.., \mathbf{r_N})d\mathbf{r}_3.d\mathbf{r}_N \quad (5.8b)$$

The expectation values of one-electron operators, like the electron density or the elastic scattering, can be calculated using Γ^1, whereas energy depends on two-electron operators and therefore requires Γ^2. In fact,

$$\rho(\mathbf{r}) = N \int \psi(\mathbf{r}_1, \mathbf{r}_2.., \mathbf{r_N}) \cdot \psi^*(\mathbf{r}_1, \mathbf{r}_2.., \mathbf{r_N})d\mathbf{r}_2..d\mathbf{r}_N = \Gamma^1(\mathbf{r}_1; \mathbf{r}_1) \quad (5.9a)$$

$$F(\mathbf{k}) = \int \psi^*(\mathbf{r}_1, \mathbf{r}_2, \ldots \mathbf{r}_n) \sum_{j=1}^n e^{i\mathbf{k}\cdot\mathbf{r}_j} \psi^*(\mathbf{r}_1, \mathbf{r}_2, \ldots \mathbf{r}_n)d\mathbf{r}_1 d\mathbf{r}_2 \ldots d\mathbf{r}_n =$$

$$= \int \rho(\mathbf{r}_1)e^{i\mathbf{k}\cdot\mathbf{r}_1}d\mathbf{r}_1 = \int \Gamma^1(\mathbf{r}_1; \mathbf{r}_1)e^{i\mathbf{k}\cdot\mathbf{r}_1}d\mathbf{r}_1 \quad (5.9b)$$

If the wavefunction is calculated with molecular orbitals ϕ_i, which are linear combinations of atomic orbitals χ_j:

$$\phi_i(\mathbf{r}) = \sum_{j=1}^m c_{ij}\chi_j(\mathbf{r}) \quad (5.10)$$

then one can define a matrix $\mathbf{P} = \mathbf{C}^\dagger \mathbf{N} \mathbf{C}$ where $\mathbf{N}$ is a diagonal matrix containing the occupation number n_i of each molecular orbital ϕ_i. With this matrix, one can easily write the one-electron reduced density matrix as:

$$\Gamma^1(\mathbf{r}_1; \mathbf{r'}_1) = Tr\left[\mathbf{P}\{\chi_{ij}(\mathbf{r})\}\right] \quad (5.11)$$

where $\{\chi_{ij}(\mathbf{r})\}$ is the matrix of all orbital products $\chi_i\chi_j$. The electron density and the scattering factor also assume simple forms:

$$\rho(\mathbf{r}) = Tr(\mathbf{P}\{\chi_{jj}(\mathbf{r})\}) \qquad (5.12a)$$

$$F(\mathbf{k}) = Tr[\mathbf{P}\{f_{ij}(\mathbf{k})\}] \qquad (5.12b)$$

where $\{f_{ij}(\mathbf{k})\}$ is the matrix of Fourier transform of all orbital products $\chi_i\chi_j$.

Given the correlation between density matrix elements and structure factors, addressed in Eq. (5.12b), it may be tempting a direct refinement of $\mathbf{P}\{\chi_{jj}(\mathbf{r})\}$ elements from X-ray diffraction measurements, like for the coefficients of the multi-pole expansion discussed in Chap. 3. However, this is not possible, because the *N-representability* of the density matrix would be lost. The *N-representability* is the condition that the density matrix must correspond to a wavefunction suitable for particles like electrons that are *fermions*, i.e. particles for which the wavefunction must by antisymmetric. For two particles, this condition is easily shown through the exchange of the two particles:

$$\psi(\mathbf{r}_1, \mathbf{r}_2) = -\psi(\mathbf{r}_2, \mathbf{r}_1) \qquad (5.13)$$

Clinton and Massa (1972) and Clinton et al. (1973) proposed some criteria to impose the idempotency condition to a density matrix and therefore circumvent the problem, being able to retrieve a $\mathbf{P}\{\chi_{jj}(\mathbf{r})\}$ from the X-ray diffraction. Using similar methodologies Tsirelson et al. (1980) derived density matrices for lithium formiate. Alexandrov et al. (1989) and Schmider, Smith and Werich (1993) went even further and coupled Bragg X-ray diffraction and Compton scattering, thus reconstructing density matrices in position and in momentum space for simple elemental solids like Silicon, Diamond (Alexandrov 1989), Beryllium and Neon (Schmide et al. 1993).

Following a different stream, Stewart (1969), Coppens et al. (1970), Coppens et al. (1971), Coppens et al. (1971) and Stewart et al. (1975) attempted the refinement of atomic wavefunction coefficients from X-ray diffraction data. However, Stewart and Coppens eventually moved toward models of the electron densities, instead of the wavefunction, as described in Chap. 3. Density models are more straightforwardly derived from experimental X-ray diffraction and gave rise to the popular multipolar model. It does not return wavefunction coefficients, unless atoms are isolated enough to be considered as non-interacting, a condition rarely realized for atoms in molecules or in crystals. For strongly covalent bonds in organic molecules and in covalent crystals, the interplay between atoms is such that atomic wavefunctions are lost in favor of molecular/crystalline ones. However, the assumption of isolated atoms could be reliable in coordination complexes where metal ions interact with coordinating ligands with low-overlap of their orbitals. Thus, the density functions refined for the metal can be a direct representation of the metal orbital functions. Holladay et al. (1983) proposed to retrieve the valence d-orbital population of transition metals, assuming the metal d-orbital do not mix with any other orbital in the system. This enables constructing a symmetric 5 x 5 density matrix, consisting of 15 independent orbital products. As discussed above, the products of spherical harmonics can be expanded in terms of linear combination of spherical harmonics themselves (Rose 1957), thus:

$$y_{l_1,m_{l_1}}\, y_{l_2,m_{l_2}} = \sum_i y_{L_i,m_{L_i}} \tag{5.14}$$

Suitable values of L_i are l_1+l_2, l_1+l_2-2, ..., $|l_1-l2|$. Thus, for a product between the angular parts of the d-orbitals, the involved spherical harmonics are the monopole ($l = 0$), the 5 quadrupoles ($l = 2$) and the 9 hexadecapoles ($l = 4$). By equalizing (5.12a) and Eq. (3.7), one can find a common expression for the electron density of nd orbitals of a metal ion:

$$\begin{aligned}
\rho_{nd\_orbital}(\mathbf{r}) &= \sum_{\nu \in nd}\sum_{\mu \in nd} P_{\nu\mu}\chi_\nu(\mathbf{r})\chi_\mu(\mathbf{r}) = \left(R_{nd}(\mathbf{r})\right)^2 \left(\sum_{\nu \in nd}\sum_{\mu \in nd} P_{\nu\mu}\, y_{2,m_\nu}(\mathbf{r}/r)y_{2,m_\mu}(\mathbf{r}/r)\right)\\
&= \left(R_{nd}(\mathbf{r})\right)^2 \left(\sum_{l=0,2,4}\sum_{m_l=-l}^{+l} P_{l,m_l}\, y_{l,m_l}(\mathbf{r}/r)\right)\\
&= R_{n,l}(r)\left(\sum_{l=0,2,4}\sum_{m_l=-l}^{+l} P_{l,m_l}\, y_{l,m_l}(\mathbf{r}/r)\right) = \rho_{nd\_multipole}(\mathbf{r})
\end{aligned} \tag{5.15}$$

$y_{2,m_\nu}(\mathbf{r}/r)$ and $y_{2,m_\mu}(\mathbf{r}/r)$ are the angular part of d-orbitals, $R_{nd}(\mathbf{r})$ and $R_{n,l}(\mathbf{r})$ are the radial part of the orbital and multipolar functions, respectively. Thus, there is a unique correspondence between the coefficients of the multipolar expansion and the orbital products, for the localized density matrix of the d-orbitals. Stevens and Coppens (1979) refined d-orbital coefficients (for specific orbital symmetries) by including them into the expression of the electron density, whereas Holladay et al. (1983) proposed a generalized calculation of d-orbital population from the refined multipoles, valid for any kind of symmetry (an anticipation of this work was published by Stevens 1980). A 15×15 matrix links the coefficients of the orbital products to the multipole coefficients. We may therefore consider the multipolar refinement of metal valence shells as a kind of "partial" wave function refinement, localized just at the metal atom. The method has been widely adopted in many studies on metal organic complexes and networks.

5.2 Modern Methods

More thorough methods have been proposed in the literature to calculate wavefunctions or density matrix elements from X-ray diffraction experiments. Some notable examples are the X-ray atomic orbital (XAO, by Tanaka et al. 2008) and its development (X-ray Molecular orbital model, Tanaka, 2018; Tanaka and Wasada-Tsutsui 2021), the valence orbital model (VOM, by Figgis, Reynolds and Williams, 1980), the Molecular Orbital Occupation Number (MOON, by Hibbs et al. 2005), and the spin-resolved atomic orbital model (Kibalin et al. 2021). In XAO, one directly refines selected electronic states, within the framework of perturbation theory. The coefficients of each orbital are refined against the observed structure factors, keeping the orthonormal relationships among the atomic orbitals. In VOM, a low overlap between ligand and metal orbitals is explicitly assumed, in keeping with the Ligand Field Theory (Figgis 1966). Although somewhat closer to a complete density matrix or wavefunction refinement from X-ray diffraction, VOM and XAO have been applied more rarely than the multipolar model. In MOON, the occupation numbers of molecular orbitals are refined. It is noteworthy that this method enables going somewhat beyond the single Slater determinant,[3] i.e. wavefunction antisymmetrized by assuming a unique electronic configuration for the molecule/crystal. In MOON, however, the coefficients of atomic orbitals are not changed during the optimization, so the only actual variables are the occupation numbers.

5.2.1 X-Ray Restrained Wavefunctions: Canonical and Extremely Localized Molecular Orbitals

A more generalized method, presently the most popular, is the X-ray restrained wave function (XRW),[4] introduced by Jayatilaka (1998), who calculated the wavefunction of metallic Be (see Fig. 5.2), and later developed with some coworkers (Jayatilaka and Grimwood 2001; Grimwood and Jayatilaka 2001; Bytheway et al. 2002; Bytheway et al. 2002; Grimwood et al. 2003) following the strategies by Clinton and Massa since the 1960s (Clinton et al. 1969; Clinton and Massa 1972; Clinton et al. 1973). XRW allows extracting a plausible wave function from experimental X-ray diffraction data

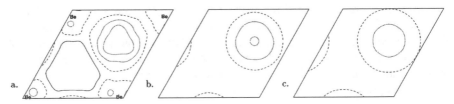

Fig. 5.2 The electron density in some planes of the crystal structure of Be. The electron density is calculated with the XRW method using X-ray diffraction data from Iversen et al. (1995). Reproduced from Jayatilaka (1998) with the permission of the American Physical Society

---

[3] A Slater Determinant is a way to consider all permutations of electrons in orbitals, while respecting the Pauli principle that for fermionic particles (like electrons) hampers the possibility of two particles simultaneously in the same quantum state. A single determinant implies that a single electronic configuration is considered. This is the assumption in Hartree-Fock method.

[4] The method is often called X-ray constrained wavefunction. However, it is inappropriate to say that there is a real constraint (as we will see in the following). See for example discussion in Ernst, Genoni and Macchi (2020).

and it can be considered as a way of merging wave function methods and density functional theory (DFT),[5] see Genoni et al. (2017).

In the XRW strategy, a crystal is simulated with N_m non-interacting and rigidly identical molecules. The unit-cell electron density is the sum of the identical molecular electron densities $\rho_k(\mathbf{r})$, obtained from a reference distribution $\rho_o(\mathbf{r})$ after applying the N_m unit-cell symmetry operations $\mathbf{R}$:

$$\rho_{cell}(\mathbf{r}) = \sum_{k=1}^{N_m} \rho_k(\mathbf{r}) = \sum_{k=1}^{N_m} \rho_o\big(\mathbf{R}_k^{-1}(\mathbf{r} - \mathbf{r}_k)\big) \tag{5.16}$$

$\mathbf{r}_k$ is the center of the kth molecule in the unit cell. $\rho_o(\mathbf{r})$ is initially approximated with the electron density calculated from a single Slater determinant wavefunction (like for the classical Hartree-Fock) that not only variationally minimizes the electronic energy, but also reproduces, within a given precision, a set of observed structure factor amplitudes. The restraint forces the global electron density (5.16) toward the "exact" electron density, while maintaining a quantum mechanical form (Jayatilaka and Grimwood 2001). This implies finding the set of Molecular Orbitals of a Slater determinant that minimize the expectation value of the functional:

$$J[\psi] = \mathcal{H}[\psi] + \lambda\big(\chi^2[\psi]\big) \tag{5.17}$$

where $\mathcal{H}$ is the Hamiltonian operator applied to a single Slater determinant molecular orbital wavefunction ψ, λ is the link to the external data (here a set of structure factors, but potentially any other observable). χ^2 measures the statistical agreement between the structure factor amplitudes calculated with the wavefunction and those supplied externally (experimentally or theoretically):

$$\chi^2[\psi] = \frac{1}{N_r - N_p} \sum_{\mathbf{k}} w_{\mathbf{k}}\big(\eta F_{\mathbf{k}}^{calc}[\psi] - F_{\mathbf{k}}^{obs}\big)^2 \tag{5.18}$$

N_r is the number of collected X-ray diffraction data, N_p is the number of adjustable parameters (thus $N_r - N_p$ represents the degrees of freedom), $\mathbf{k}$ is the scattering vector, $w_{\mathbf{k}}$ is the weight of the reflection $\mathbf{k}$ (often taken as $1/\sigma_{\mathbf{k}}^2$, where $\sigma_{\mathbf{k}}$ is the standard uncertainty of the observed structure factor amplitude $F_{\mathbf{k}}^{obs}$) and η is the scale factor. In principle, the only refined parameter is η, however more parameters are effectively adjusted during this procedure: (a) λ is progressively increased until χ^2 becomes smaller than a desired threshold; (b) the molecular orbital coefficients

---

[5] Density functional theory (DFT) is a method of solving Schrödinger equation that does not rely on the wavefunction but on the electron density and expresses energy (or any other property of a system) as a functional of the electron density. It originates from the famous theorems by Hohenberg–Kohn: (1) The total energy is a unique functional of the electron density; (2) The functional that delivers the ground-state energy of the system gives the lowest energy if and only if the input density is the true ground-state density (Hohenberg and Kohn, 1963).

are modified at each cycle by minimizing the expectation value of $J[\psi]$. For this reason, the meaning of χ^2 is ambiguous, and the desired value of 1.0 (the target of a typical crystal structure refinement) does not guarantee the best model.

The currently adopted procedure has some pitfalls:

(a)   The wavefunction is calculated for a molecule, but its electron density is constrained to reproduce structure factor amplitude of a crystal. Therefore, $\mathcal{H}[\psi]$ and $\chi^2[\psi]$ do not represent the same realm and may be in contrast with each other.

(b)   The functional $J[\psi]$ has unbalanced units (energy on the left, unit-less on the right, if the weight is $1/\sigma_k^2$, or square charge if the weight is unitary). This implies that J cannot not correspond to any physical observable and that the same value of λ affects in a different way the wavefunctions of crystals with different number of electrons and different volumes. Moreover, while $\mathcal{H}[\psi]$ returns a total energy of the system, χ^2 is a square difference between Fourier transformed charge distributions.

(c)   The electronic state of the system (for example the spin state) is fixed by the Hamiltonian part of $J[\psi]$ and cannot be modified by the restraint to the experiment.

Some recent work (Genoni et al. 2017) demonstrates that the best convergence between highly correlated theoretical calculations and wavefunction refinements against the structure factors calculated at this high level of theory occur at intermediate resolution ($\sin\theta/\lambda \sim 0.5$–$0.7 \text{Å}^{-1}$) instead of the highest resolution, which seems somewhat surprising. Similar conclusions hold for the incorporation in the final electron density of the effects of the crystal field (Ernst et al. 2020).

After the successful applications of this method, many similar approaches have been proposed and tested. One of the most popular is the one that restrains to X-ray intensities a molecular orbital wavefunction calculated from extremely localized molecular orbitals (ELMOs), see Genoni (2013a, b). In this procedure, the molecular orbitals are strictly localized within predefined fragment of the molecule/polymer in the crystal and the restrained to reproduce the X-rays diffracted intensities. Although the constraint (the localization) produces a wavefunction less suitable than an unconstrained wavefunction, there are some advantages:

(a)   The fragments allow devising databanks like for the transferable electron density models based on multipoles (see Sect. 3.5);

(b)   The molecular orbitals can be interpreted in chemically sensible terms (see for example, Ernst et al. 2020);

(c)   A valence bond treatment is possible, which is also useful for a chemical interpretation, see for example (Genoni 2017; Casati et al. 2017).

5.2.2 Combined Molecular Orbital Calculations and Structural Solution and Refinement

In recent years, the interplay between X-ray diffraction and solid-state quantum chemical methods has increased, especially because of the terrific increase of structure solutions of molecular crystals from powder X-ray diffraction. Techniques under the name of "DFT—Assisted crystal structure solutions" (Gautier et al. 2013) or First Principle Assisted Structure Solutions (Michel et al. 2019) have been ever more systemically employed to solve difficult molecular or polymeric structures from powder X-ray diffraction data, notoriously more difficult than single crystal diffraction. These methods use a blend of classical Rietveld refinement[6] (Rietveld, 1969) and minimization of energies with DFT calculations. This joint venture reduces the human intervention in sorting out troubles of Rietveld-based models because molecular conformations are automatically corrected by the DFT calculation during the exploration of the parameter space.

Theoretical methods to predict crystal structures are also very popular and may assist the structure solution attempts, but they won't be discussed in this section. The reader is invited to consider specialized literature (see or example, Price, 2018).

A step forward in this series of methods concerns the possibility to use molecular quantum chemical calculations to improve the crystal structure model, making use of techniques derived from the X-ray restrained wavefunction approach described in the previous paragraph. Among these techniques, the Hirshfeld Atom Refinement (HAR) is the most well-known. The method is named after F. L. Hirshfeld, who proposed a partition scheme very simple but surprisingly effective (Hirshfeld 1977b), the *stockholder* partition, which treats the electron density as a "financial" function and the atomic ground state electron density as the initial "investment" of an atom. At each point $\mathbf{r}$, the electron density of an atom j is a part of the total electron density $\rho(\mathbf{r})$ of a molecule, proportional to the contribution of the atom to the *promolecule* density:

$$\rho_j(\mathbf{r}) = \left[\frac{\rho_j^{IAM}(\mathbf{r})}{\sum_i \rho_i^{IAM}(\mathbf{r})} \right] \rho(\mathbf{r}) \tag{5.19}$$

HAR was proposed by Jayatilaka and Dittrich (2008), who proposed to partition a theoretical molecular electron density into atomic densities $\rho_j(\mathbf{r})$, and derive aspherical form factors. At variance from the multipolar model, no parameter of the atomic electron densities is refined, but only the atomic positions and displacements. Therefore, in HAR the electron density is purely theoretical, but based on a molecular geometry which is refined against experimental data.

The advantage of this method stands in the improved quality of the structural model, which becomes especially relevant when H atoms are concerned. Applications of HAR (Capelli et al. 2014; Fugel et al. 2018), demonstrated that molecular

---

[6] The Rietveld method is a technique to refine a structural model from powder diffraction data.

geometries can be very accurate. In particular, HAR avoids too short X–H distances typically refined with classical IAM models.

HAR is conceptually similar to the generalized X-ray scattering factor (GSF) approach (Stewart 1969, 1973; Stewart et al. 1975). As a matter of facts, GSF refinements return atomic positions of H atoms in close agreement with models refined against neutron diffraction (Destro et al. 2000). In Stewart's approximation, the atomic partition of the form factors is based on the projection of the molecular electron density onto a set of atomic Legendre polynomials. In the practical applications (Destro et al. 2000), only the H form factor is used (all other atoms come from a IAM). The H form factor does not depend on the geometry of the molecule, but it comes from pre-calculated X-H densities adopting standard bond distances (Stewart et al. 1975). Therefore, the real improvement of HAR is the recursive type of the calculation based on the actual molecule, an approach that would be technically possible also with GSF.

HAR has also been coupled with extremely localized molecular orbital calculations (HAR-ELMO, Malaspina et al. 2019), offering the opportunity to reduce the computational costs if a set of calculated libraries of ELMO's is available.

An extension of the XRW (in combination with HAR), is the possibility to refine with a least square procedure the typical structural parameters of conventional refinements (atomic position and displacements) together with the calculation of the wavefunction. As discussed in Sect. 5.2.2, the wavefunction is not parametrized in the same way as the multipolar model, so that some (few) coefficients can be freely refined against the measured structure factors. A necessary step is to adopt a Hirshfeld atom partition of the electron density calculated at a XRW cycle (or any other kind of subdivision of the electron density in atomic terms) and transform the molecular orbital wavefunction into a series of atom centered scattering factors. These could be used for refining the atomic positions and displacement parameters. The procedure can be cycled recursively until convergence is obtained. The procedure was called *X-ray wavefunction refinement* (Grabowsky et al. 2012). Noteworthy, a true wavefunction refinement was the attempt by Coppens, Csonka and Willoughby (1970), who tried in fact to refine directly molecular orbital coefficients. The procedure by Grabowsky et al. (2012) is instead a structure refinement using the XRW.

5.2.3   Reduced Density Matrices Refinement

In alternative to wavefunction calculations restrained to elastic diffraction, the combination of elastic Bragg diffraction and inelastic Compton scattering provides information that enables refinement of elements of the one-electron reduced density matrix (Gillet 2007). Elements of the density matrix are: (a) diagonal terms, which give the electron density, hence connected with Bragg diffraction; (b) off-diagonal terms, that provide the autocorrelation of the wavefunction, hence connected with the Compton scattering. The Compton effect originates from the inelastic scattering of a photon

Table 5.1 The relationship between various electron density functions in momentum ($\mathbf{p}$) and position ($\mathbf{r}$) space (Cooper 1985). $\psi(\mathbf{r})$ and $\chi(\mathbf{p})$ are the one-electron wavefunction. $P(\mathbf{r})$ is the *pair correlation function* of the electron density in position space (better known as the *Patterson function*), $P(\mathbf{p})$ is the *pair correlation function* of the electron density in momentum space; $B(\mathbf{r})$ is the *reciprocal form factor*; $F(\mathbf{p})$ is the *static form factor*. The symbols * and ^2 indicate the operations of *autocorrelation* or squaring, respectively

Momentum space	Fourier transform	Position space
$F^2(\mathbf{p})$	$\leftrightarrow$	$P(\mathbf{r}) = \int \rho(\mathbf{r}')\rho(\mathbf{r}' - \mathbf{r})d\mathbf{r}'$
↑ ^2		↑ *
$F(\mathbf{p}) = \int \psi(\mathbf{p}')\psi(\mathbf{p}' - \mathbf{p})d\mathbf{p}'$	$\leftrightarrow$	$\rho(\mathbf{r}) = |\psi(\mathbf{r})|^2$
↑ *		↑ ^2
$\psi(\mathbf{p})$	$\leftrightarrow$	$\psi(\mathbf{r})$
↓ ^2		↓ *
$\pi(\mathbf{p}) = |\psi(\mathbf{p})|^2$	$\leftrightarrow$	$B(\mathbf{r}) = \int \psi(\mathbf{r}')\psi(\mathbf{r}' - \mathbf{r})d\mathbf{r}'$
↓ *		↓ ^2
$P(\mathbf{p}) = \int \pi(\mathbf{p}')\pi(\mathbf{p}' - \mathbf{p})d\mathbf{p}'$	$\leftrightarrow$	$B^2(\mathbf{r})$

off a free electron. The applications of Compton scattering were developed by Weiss (1966) and then by Weyrich (1996) and by Cooper et al. (2004).

With modern synchrotron radiation sources, it is nowadays possible to measure precise Compton profiles up to sufficiently large scattering vector moduli and therefore refine pseudoatom models, inspired by the multipolar models described in Chap. 3 (see for example, Gillet et al. 2001; Gillet 2007). Through the refinement of the density matrix, the connection between electron density distribution in position space, $\rho(\mathbf{r})$, and in momentum space, $\pi(\mathbf{p})$, would be fully established because the functions are linked through precise transformation as described in Table 5.1. However, electron distribution in momentum space is far less popular than in position space. As explained by Cooper (1985):

> It appears wholly natural to think and work in terms of position space densities, a switch to the momentum representation seems perverse and hardly justified by the mere fact that momentum densities are accessible to measurement through the Compton effect. Only in delocalised systems such as electrons in metals is there any hint that a description in terms of momentum, *p*, or at least wavevector, *k*, might be preferable. Yet by adopting the complementary viewpoint a real insight can be gleaned into the behaviour of electrons in isolated atoms and molecules as well as in condensed matter.

The work by Gillet (2007) anyway paved the way for a unification of the field that could eventually result in a joint refinement of electron charge and electron spin density (or wavefunction), represented both in momentum and position space. Noteworthy, while experimentally one cannot measure simultaneously an electron position and momentum, as discussed above, nothing hampers the possibility of reconstructing the corresponding densities from a combination of different experiments.

This goal is however not reached yet, but it could be an objective for the next years of research.

Chapter 6
Quantum Crystallography and the Standard Model

The combination of quantum theory and special theory of relativity forms the *quantum electrodynamics*. This theory considers only one of the four fundamental forces, namely the electromagnetic force (albeit being itself the unification of electric and magnetic forces). If electromagnetism is coupled with the nuclear weak force, we obtain the *electroweak theory* and the further combination with *quantum chromodynamics* (the theory of the strong nuclear force) gives the *grand unified theory*, whereas it is well known that a unification with gravitation has not been reached yet. The particles introduced by these theories and their interactions is the so called *Standard Model* (Griffiths 1987; for an historical overview, see Weinberg 2004). A question for the study of quantum crystallography is how much these theories affect the models and what can be the role of crystallography.

6.1 Relativity

Another extraordinary revolution accompanied the advent of the quantum theory at the beginning of the XX century: the theory of relativity, introduced first in a restricted (Einstein 1905) and then in a generalized treatment (Einstein 1916). In the special relativity, there are two postulates: the physical laws are invariant with respect to the reference system and the speed of light in vacuum (c) is a constant for all observers despite their relative motion. One important change due to the relativity is the concept of mass at rest, being different from the mass at the speed of the particle. This implies a correction to the kinetic energy of a system:

$$E = \sqrt{c^2 \mathbf{p} \cdot \mathbf{p} + \left(mc^2\right)^2} \tag{6.1}$$

where **p** is the momentum of the particle and m is its mass at rest. This implies corrections to the standard quantum mechanical Hamiltonian adopted in the non-relativistic treatment. However, this generates many complications in practice that lead to many possible schemes to solve the Schrödinger equation within a relativistic treatment, for example the Klein-Gordon scheme (Klein 1926; Gordon 1926). Here, we focus on the main consequences of relativity for crystallography.

Despite initially underestimated by Dirac himself (Dirac 1929), there are some important consequences for atoms due to relativity:

(a)   the *relativistic contraction*: it is due to the increase of the relativistic mass $m_{e,rel}$ of an electron with respect to the mass at rest m_e:

$$m_{e,rel} = \frac{m_e}{\sqrt{\left(1 - \frac{v_e}{c}\right)}} \tag{6.2}$$

where v_e is the speed of the electron (obviously $v_e < c$). This implies a smaller Bohr radius for the electron and consequently a smaller size of the atom. The stabilization of atomic s and p states is stronger for electrons near the atomic nucleus (because they move faster) and therefore not negligible for atoms with large atomic number, because their core electrons are more contracted toward the highly positive nucleus. In structural chemistry, the main consequence is a surprising reduction of the size of atoms of lower rows in the periodic table. For quantum crystallography, the main effect concerns the atomic form factors, which are larger than expected from a non-relativistic treatment at high diffraction angles for heavier elements. See for example Fig. 6.1, where uranium form factors calculated with or without relativistic treatment are compared.

The relativistic behavior may affect the modelling of core electron densities, especially for heavier elements, unless special attention is paid to select appropriate atomic wavefunctions to compute the form factors of core electrons.

(b)   The *spin–orbit* coupling: the coupling between the two kinds of angular momentums of an electron (the one due to the orbital and the one due to the spin) is correctly evaluated only in a relativistic treatment. This has consequences, especially when considering the magnetization of an atom and the orientation of the magnetic moment with respect to the crystal axes. This also affects the exact determination of the spin electron density.

(c)   *Valence shell stabilizations* or *destabilizations*: despite the most important effects concerns the core electrons, stabilization of s and p atomic orbitals, at the expense of d and f, occurs also for external shells. One of the most interesting roles played by the relativistic behavior of valence electrons is the surprising attraction between closed shell d^{10} metal ions, also known as *metallophilicity* (Pyykko and Desclaux 1979), which is in fact lower for Cu and larger for Au.

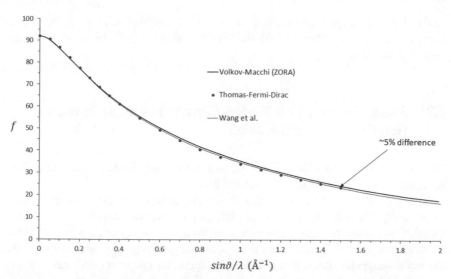

Fig. 6.1 The atomic form factor of uranium, computed with two kinds of non-relativistic wave-functions and compared with a relativistic wavefunction. Thomas and Umeda (1956) calculated the form factor using the Thomas–Fermi-Dirac non-relativistic method (blue circles). Wang et al. (1993) using Roothan-Hartree–Fock non-relativistic method with Clementi and Roetti (1974) and McLean and McLean (1981) wavefunctions (red line). Volkov and Macchi (black line) calculated the atomic form factor using ZORA and the Perdew, Burke and Ernzerhof (1996) exchange–correlation functional with high quality basis sets, obtaining results remarkably close to Coulthard (1967) (not shown in the graph and differing for less than 0.1% at all points); these data are included in the XD2006 software (Volkov et al. 2006). The discrepancy between relativistic and non-relativistic starts at $sin\vartheta/\lambda \sim 0.5\,\text{Å}^{-1}$ and it is significant (up to ca. 5%) for $sin\vartheta/\lambda > 1.0\,\text{Å}^{-1}$

Relativistic corrections are considered for atomic core electron densities in multi-polar model. For example, Su and Coppens (1998) and Macchi and Coppens (2001) fitted atomic wavefunctions obtained with non-relativistic atomic calculations to the numerical solution of Dirac–Fock fully relativistic calculations. These functions describe better the radial behavior of atomic core electron densities, while preserving the same easy implementation in the standard multipolar formalisms described in Chap. 3. Later, Volkov and Macchi (included in Volkov et al. 2006) calculated the atomic form factors using the zero-order regular approximation (ZORA) and a DFT functional.

Relativistic treatments have also been included in the standard XRW approach. The flexibility of the XRW scheme, in fact, enables in principle using all kinds of quantum chemical methods, not only the Hartree–Fock or the DFT procedures, but also the post-Hartree–Fock methods (describing, at least in part, the electron correlation which is missing in Hartree–Fock). In this respect, the non-relativistic Hamiltonians can be replaced by so-called Coulomb-Dirac Hamiltonians that calculate the correct kinetic energy and consider a properly adjusted wavefunction as a four-component vector able to reproduce the scalar relativistic contraction and the vecotrial spin–orbit coupling in the proper way. Bučinský, Jayatilaka and Grabowsky

(2016) have first applied this method to crystals of diphenyl mercury and triphenyl bismuth, showing the effects of relativistic correction on QTAIMC analysis of the chemical bonding.

6.2 Going Too Far: Can We See Charge-Parity Symmetry Violation with Diffraction?

For the reader who understands the question before any further explanation, it is necessary to give immediately the answer: NO!

Though, let's see the reason for this short paragraph concluding Chapter 6. Apart from gravity, the standard model of particle physics includes the other three fundamental forces (electromagnetic, weak nuclear force and strong nuclear force).[1] From what we have seen so far, it is clear that quantum crystallography deals mainly with the electromagnetic force, which is true for quantum chemistry and more generally for chemistry as well (apart from nuclear chemistry). Nonetheless, experiments involving for example neutron radiation do depend also on the (residual) strong nuclear forces. Crystallographers take them as granted in the evaluation of the atomic scattering lengths of neutron radiation and do not further test or validate with new experiments. This differs from what happens to electron scattering factors that are constantly improved with new models as we have seen through this review.

The weak force is normally not considered in crystallography. The name weak is in fact quite indicative. The weak force is much weaker than the strong nuclear force, and anyway weaker than the electromagnetic. Nevertheless, the weak force is responsible for one of the most intriguing phenomenon in physics, which is the charge symmetry / parity symmetry violation, also known as CP-violation (Christenson et al., 1964). The violation implies that the inversion of charge and symmetry does imply a change in the physics. If we ignore the weak forces, the inversion of charge would not produce any change in the physics of an atom, a molecule or a crystal. An *anticrystal* made of *antimatter* atoms would be just identical to a crystal. Instead of having positive nuclei and negative clouds around them, we would have negative nuclei and positive clouds, but the energy of the quantum states would not change, so the involved chemistry. In structural chemistry and crystallography, the CP-symmetry has also the consequence that two enantiomeric forms of the same molecule or crystal have the same energy.

What happens, though, if we consider the weak forces and thus we consider the possible effects of CP-violation? For chiral species Salam (1991, 1992) suggested that a second-order phase transition below a critical temperature may be the reason of the natural abundance of enantiomers of natural biomolecules, in particular amino acids. Although not fully proved, CP-violation could be the cause of this transformation.

---

[1] The strong nuclear force mainly acts between the quarks forming protons or neutrons. The force attracting neutrons to protons in nuclei is the *residual strong force*.

Why are we concerned about this? Recently, Belo et al. (2018) reported the different crystallographic behavior between crystals of L-Alanine and D-Alanine, i.e. the two enantiomorphs of one of the most important natural amino acids. Differences concerned the temperature evolution of the two crystal phases. Because a different structural behavior implies a different energy involved (apparently confirmed by calculations as well), this experiment would be a seamlessly spectacular confirmation of Salam's hypothesis. However, as pointed out by Bürgi and Macchi (2018), the observations can be explained with different treatments of the samples, as most of the data were measured in different conditions and with different "crystal histories". Noteworthy, the theoretical calculations (of quantum mechanical fashion) cannot return any effect of the CP-violation because they do not include the weak forces. Thus, necessarily, the theoretical results must confirm the CP-symmetry. Moreover, as pointed out by Quack, Stohner and Willeke (2008), the parity-violation energy difference is of the order of 10^{-11} J mol-1, a quantity that is difficult to measure unless with high resolution spectroscopy and too small to be influent in crystal structure diffraction measurements. This does not mean of course that the Salam's hypothesis is not valid, but it means that it is very difficult to prove it with diffraction experiments.

Very likely, the limit of quantum crystallography (measurements and simulations) lies at the level of the relativistic effects discussed in Sect. 6.1. Going beyond is too ambitious, at present.

Chapter 7
Quantum Crystallography in the Everyday Life of a Crystallographer and a Quantum Chemist

There are several overlapping areas between quantum crystallography and all other crystallographic fields as well as with molecular quantum chemistry and solid-state quantum physics. In this chapter, we briefly highlight these connections. Cooperation between different fields and mutual validation of results, theories, and models may take place and further develop in the future.

7.1 Experimental Crystal Structure Determination

A vast majority of crystallographers take advantage of diffraction techniques to determine crystal structures and very often the purpose is obtaining experimental molecular geometries with a high degree of accuracy or determining molecular conformations and configurations. Nevertheless, a growing branch is interested in the interaction among molecules in order to investigate the supramolecular packing.

How does quantum crystallography interplay with these important streams of scientific research? In many aspects, but only few of them have been exploited so far. We can summarize the interplay as it follows:

(a)   The diffraction experiment is the measurement of a quantum mechanical observable like the crystal structure factors, through which an approximate crystal electron density can be calculated. We call this *structure solution*. This is of course true for many analytical techniques in chemistry (all spectroscopic methods, for example), but not for all of them (e. g. chromatography, mass spectrometry, traditional analytical techniques, etc.). Few crystallographers appreciate this; however, the contribution of quantum mechanics is essential to achieve the expected results.

P. Macchi, *Quantum Crystallography: Expectations vs Reality*,
SpringerBriefs in Crystallography, https://doi.org/10.1007/978-3-030-95641-7_7

(b)   The structural model is refined by means of an atomic expansion. Even though most of the structures are solved and refined using the spherical atom model, thus a poor approximation of the exact electron density, the atomic form factors (as we saw in Chap. 2) are calculated using quantum mechanical methods, otherwise the structural models would be quite deficient (and structure solution itself may be problematic).

(c)   Advanced models for structure refinement are nowadays available in the main software adopted for ordinary crystal structure determination. Among them, we can mention the Hirshfeld Atom Refinement and the Transferable Aspherical Atomic Models. It should not escape attention that all these improved models, although requiring an advanced knowledge, can provide not only a (better) model of the crystal structure, but also insight in the chemical bonding and in the supramolecular interaction, thus enhancing the information typically extracted from a normal diffraction experiment. This implies that without additional experimental costs, the traditional outcome of a diffraction experiment may be significantly enriched. A major effort will be making appropriate use of this extra outcome.

7.2   Accurate Measurements of Crystal Properties

Crystallography is not just X-ray diffraction. The same holds true for quantum crystallography. There are many properties that characterize crystalline materials, like optical, electronic, magnetic etc., and they constitute the main objective of research in materials science. Because quantum crystallography deals with the quantum mechanical behavior of crystals, the interplay between these disciplines is straightforward. In this book, we have not considered in details complementary techniques, but it is fair to mention at list a few of them, which have provided some valuable information for quantum crystallographic research:

(a)   *Atomic Force Microscopy* and *Scanning Tunneling Microscopy*. These techniques investigate mainly the surface of crystals (as well as of non-crystalline materials), probing the van der Waals forces and the electrostatic forces or mapping the electronic density of states. The measurements provide information on the electronic structure of atoms or molecules forming the materials surface or lying on them (see for example Welker and Giessibl 2012). Conversely, quantum crystallographic models provide useful tools for a better interpretation of the observations.

(b)   *Electron diffraction.* While electron microscopes are well known and frequently used in crystallography (as well as in many other disciplines), electron diffraction has been, so far, a niche technique, reserved for fewer applications. The development of cryo-electron microscopy and of standalone electron diffractometers has resumed the interest for this technique, with enormous implications for quantum crystallography as well. The application of convergent electron beam diffraction in quantum crystallography was pioneered by Zuo et al. (1999) and by Streltsov, Nakashima and Johnson (2001). Recent advances enable using precession methods and refine models like with X-ray diffraction. Indeed, refinements based on TAAM have been already proposed (Gruza et al. 2020) which may significantly contribute to further popularization of this technique.

(c) *Electronic and optoelectronic properties.* Refraction, reflection, transmission of light, as well as generation of second or third harmonics and other non-linear optical properties are all measurements providing significant information about the polarizabilities and hyperpolarizabilities of crystals (and of their constituting molecules/building blocks). Dielectric constant, electrical conductivity, Seebeck effect, etc. are other measurements providing information about transport properties in crystals. Again, the quantum crystallographic modeling is a valuable way to enhance the knowledge of a crystalline materials obtained with those techniques, see for example Scatena, Guntern and Macchi (2019).

(d) *Magnetic properties.* Apart from neutron scattering techniques (discussed in Chap. 3), there are other techniques (like superconducting quantum interference, muon spectroscopy, etc.) that provide observables, like magnetic susceptibility and magnetization. Quantum crystallographic simulations and models are vital for correct interpretation of the results. At the same time, these measurements enable refinement of models for the magnetization and the magnetic coupling, which are of fundamental importance for the quantum crystallographic studies.

(e) *Electro-mechanic properties.* These effects are due to the interactions between electromagnetic fields and solids. Among these, piezoelectricity, electrostrictions, magnetostrictions, and radiation pressure may provide useful information, complementary with that obtained from scattering techniques.

7.3 Validating Quantum Models with Experiments

In contrast with the enormous exchange and interaction witnessed in the early days of quantum mechanics and modern crystallography, the interest in validating a theory with quantum crystallographic experiments has significantly decreased. Today, the interplay of theoretical quantum chemists with crystallography is normally limited to a bunch of molecular or crystal geometries used for validating predictions or screening the best performing theoretical models and algorithms. This is certainly regretful, because the source of information available from quantum crystallographic experiments go well beyond the simple geometrical details of a crystal structure. An example comes from the X-ray restrained wavefunction approach, a method that may foster the development of new functionals for DFT, but has not been sufficiently considered so far. Functionals are often adjusted in order to reproduce some experimental observations with greater accuracy (mostly just the geometry), whereas the XRW technique would enable devising functionals, which incorporate the experiment instead of simply using a model-mediated outcome of the experiment as calibrant.

Although one may not expect the same impact of some spectacular validations occurred in the past like, for example, the dual nature of electrons proved through electron diffraction experiments, the hope of the author is that diffraction and other crystallographic experiments be considered more systematically as potential validation of quantum mechanical theories.

Chapter 8
Concluding Remarks

8.1 The Balance Between Expectations and Reality

Throughout this Springer Brief, we considered some fundamental questions concerning the field of quantum crystallography and analyzed in detail what can be studied and what is still not possible.

Quantum mechanics is the branch of physics that deals with the sub-microscopic world, where we witness a collapse of the laws of classical physics, which are valid at the macroscopic scale. Thus, even the observation of small particles is not guaranteed. Nevertheless, the question *"Can we see the electrons?"* receives a strong *Yes*. X ray diffraction is not the only technique able to provide such information, but certainly one of the most accurate and widely adopted. Moreover, by freezing atoms in crystals, one obtains a unique opportunity to see relatively steady electron distributions, amplifying the signal by means of the periodic order. The *electron density* is the observable that allows us to say that we can see the electrons. Albeit not an observable—strictly speaking- also a crystal *wavefunction* can be computed from experiments.

A fundamental concept in quantum chemistry, although not a quantum mechanical observable, is the *chemical bond*. The question *"Can we see chemical bonds?"* is inherently inappropriate. Of course, we cannot observe a concept. Nonetheless, we are able to measure many features, often correlated with the electron density, that are genuine clues of chemical bonds. At the same time, these observables, enable unveiling the very nature of the chemical bonds, which represents the most predictive part of quantum chemistry in general, and quantum crystallography, in particular.

What are the current limits of quantum crystallography?

There are some technical limitations that likely will be overcome in the next future and some more fundamental issues that require much more research.

Among the technical limitations, we can list:

(a) X-ray restrained wavefunctions (Chap. 5) are normally calculated using a molecular approximation, but nothing hampers the calculation of truly periodic wavefunction, thus including in the Hamiltonian part a periodic potential. Only few exceptions are known till date (see Wall 2016).

(b)   While joint refinements of charge and spin density are possible thanks to combination of polarized neutron and X-ray diffraction (see Chap. 3), the combination of Bragg and Compton scattering is more complicated and so far not achieved.

(c)   Refinements of orbitals or density models (Chaps. 3 and 5) should comply with physical constraints, like the Hellmann–Feynman theorem, but only few attempts have been reported (Hirshfeld 1984).

The fundamental limitations concern instead the combination of quantum crystallographic models with experimental sources different from diffraction techniques as well as the possibility to retrieve more quantum mechanical information from the scattering techniques. At present, there is no simultaneous calculation of a quantum mechanical function (wavefunction, electron charge or spin density, electric potential, etc.) combining a scattering technique with a spectroscopic or a microscopic technique. The combination of spectroscopic measurements, like nuclear quadrupole resonance and X-ray diffraction, was envisaged long time ago by Schwarzenbach (1966) but not much put into practice thereinafter.

Albeit not discussed in this review, atomic force or scanning tunnel microscopy and magnetization measurements provide valuable quantum crystallographic information. Combining the two kinds of experiments is not impossible, but it would require significant development of the theoretical framework to devise suitable models, able to grasp information from all these sources. The advantage would be quite important, like that of mapping simultaneously bulk and surfaces or that of improving the experimental magnetization density.

As for extracting more information from scattering techniques, electron diffraction and magnetic X-ray scattering are very promising, though again there is an urgent necessity to improve the theoretical framework in order to be able to exhaust the additional information available. Finally, the impressive momentum of the free electron laser sources could guarantee improvements in quantum crystallographic dynamics. However, time dependent models are lacking at present.

8.2   Further Reading

This pamphlet has summarized only a small part of the research in the emerging field of Quantum Crystallography. It is not a textbook where to learn the theories and be acquainted with the techniques, nor a detailed report of the scientific achievements. The reader may find the basics of charge density analysis in some seminal textbooks (Tsirelson and Ozerof 1995; Coppens 1997) and review articles (Coppens and Koritzsansky 2001; Gatti 2005). Forthcoming textbooks dedicated on quantum crystallography (Macchi 2022; Massa and Matta 2021) broaden the scope and are more up to date.

Concerning the research results, some books and journal special issues reported state of the art achievements that retrospectively can tell more about the development of the field. A special issue of the Israel Journal of Chemistry was published in 1977 (issue 2-3) dedicated to electron Density Mapping in Molecules and Crystals and edited by F. L. Hirshfeld. This special issue (consisting of 19 articles and 2 appendixes) contains the fundamentals of the quantum theory of scattering, the theoretical foundations of the multipolar models, the practical aspects of multipolar refinements, and some attempts of space partitioning of the charge density. The first multi-author book on the subject (*Electron and Magnetization Densities in Molecules and Crystals*) was edited by P. J. Becker (1980). The book contains many fundamental developments (including for example the spin density determination) compared with the previous special issue as well as the anticipation of some new methods.

Jeffrey and Piniella (1991) edited another seminal book (*Application of Charge Density Research to Chemistry and Drug Design*) after a NATO-ASI meeting held in 1990. The book collects several contributions from well-known scientists active in the field, which witnessed one of the periods with the strongest momentum in the history of charge density analysis. In this book, for example, the first applications of QTAIMC to experimentally determined electron densities was proposed.

After the third edition of the European Charge Density Meeting jointly organized with a European Science Foundation Exploratory Workshop (held in Sandbjerg, Denmark, in 2003), a special issue of Acta Crystallographica Section A was published (Vol. A260, issue nr. 5, 2004), which contained many new experimental approaches and technological improvements. Inspired by the fifth edition of the European Charge Density Meeting (held in Gravdeona, Italy, in 2008), a book on *Modern Charge Density Analysis* was published (Gatti and Macchi 2012), which contained state of the art of both theoretical and experimental methods.

Recently, the first special issue explicitly dedicated to quantum crystallography was published in Acta Crystallographica B. It is a virtual special issue, collecting articles published in issues 3-6 of 2021, edited by Gillet and Macchi.

References

Abramov YA (1997) On the possibility of kinetic energy density evaluation from the experimental electron-density distribution. Acta Cryst A53:264–272

Aleksandrov YV, Tsirelson VG, Reznik IM, Ozerov RP (1989) The crystal electron energy and compton profile calculations from X-ray diffraction data. Phys Stat Sol B155:201–207

Armstrong HE (1927) Poor common salt! Nature 120:478

Aronica C, Jeanneau E, El Moll H, Luneau D, Gillon B, Goujon A, Cousson A, Carvajal MA, Robert V (2007) Ferromagnetic interaction in an asymmetric end-to-end azido double-bridged copper(ii) dinuclear complex: a combined structure, magnetic, polarized neutron diffraction and theoretical study. Chem Eur J 13:3666–3674

Bader RFW (1990) Atoms in molecules: a quantum theory. In: International series of monographs on chemistry 22, Oxford Science Publications, Oxford.

Bader RFW (2001) The zero-flux surface and the topological and quantum definitions of an atom in a molecule. Theor Chem Acc 105:276–283

Becke AD, Edgecombe KE (1990) A simple measure of electron localization in atomic and molecular systems. J Chem Phys 92:5397–5403

Becker PJ (1980) Electron and magnetization densities in molecules and crystals. Springer, Boston

Becker PJ, Coppens P (1985) About the simultaneous interpretation of charge and spin density data. Acta Cryst A41:177–182

Belo EA, Pereira JEM, Freire PTC, Argyriou DN, Eckert J, Bordallo HN (2018) Hydrogen bonds in crystalline D-alanine: diffraction and spectroscopic evidence for differences between enantiomers. IUCrJ 5:6–12

Bentley J, Stewart RF (1974) Core deformation studies by coherent X-ray scattering. Acta Cryst A30:60–67

Berlin T (1951) Binding regions in diatomic molecules. J Phys Chem 19:208–213

Bethe H (1929) Termaufspaltung in Kristallen. Ann Physik 395:133–208

Blanco MA, Martín Pendás Á, Francisco E (2005) Interacting quantum atoms: a correlated energy decomposition scheme based on the quantum theory of atoms in molecules. J Chem Theory Comput 1:1096–1109

Bloch F (1928) Über die Quantenmechanik der Elektronen in Kristallgittern. Z Physik 52:555–600

Bohr N (1920) Über die Serienspektra der Elemente. Z Phys 2:423–478

Boucherle JX, Gillon B, Maruani J, Schweizer J (1982) Determination by polarized neutron diffraction of the spin density distribution in a non-centrosymmetrical crystal of DPPH:C_6H_6. J De Physique Colloque C7(43):227–234

Bouhmaida N, Ghermani N-E, Lecomte C, Thalal A (1997) Modelling electrostatic potential from experimentally determined charge densities II. Total potential. Acta Cryst A53:556–563

Bouhmaida N, Dutheil M, Ghermani NE, Becker P (2002) Gradient vector field and properties of the experimental electrostatic potential: application to ibuprofen drug molecule. J Chem Phys 116:6196–6204

Bragg WH, Bragg WL (1913) The reflexion of X-rays by crystals. Proc R Soc Lond A 88:428–438

Bragg WL, James RW, Bosanquet CH (1922) The distributions of electrons around the nucleus in the sodium and chlorine atoms. Phil Mag 44:433–449

Bragg WL (1927) The structure of silicates. Nature 120:410–414

Brock CP, Dunitz JD, Hirshfeld FL (1991) Transferability of deformation densities among related molecules: atomic multipole parameters from perylene for improved estimation of molecular vibrations in naphthalene and anthracene. Acta Cryst B47:789–797

Brown PJ (1992) Magnetic scattering of neutrons. In: Wilson AJC (ed) International tables for crystallography, vol C. Kluwer, Dordrecht, pp 512–514

Brown PJ, Capiomont A, Gillon B, Schweizer J (1979) Spin densities in free radicals. J Mag Mag Mater 14:289–294

Bučinský L, Jayatilaka D, Grabowsky S (2016) Importance of relativistic effects and electron correlation in structure factors and electron density of diphenyl mercury and triphenyl bismuth. J Phys Chem A 120:6650–6669

Bürgi HB, Capelli SC, Goeta AE, Howard JAK, Spackman MA, Yufit DS (2002) Electron distribution and molecular motion in crystalline benzene: an accurate experimental study combining CCD x-ray data on C_6H_6 with multitemperature neutron-diffraction results on C_6D_6.

Bürgi HB, Dunitz JD (1983) From crystal statics to chemical dynamics. Acc Chem Res 16:153–161

Bürgi HB, Macchi P (2018) Comments on 'Hydrogen bonds in crystalline D-alanine: diffraction and spectroscopic evidence for differences between enantiomers'. IUCrJ 5:654–657

Bytheway I, Grimwood DJ, Figgis BN, Chandler SC, Jayatilaka D (2002) Wavefunctions derived from experiment. IV. Investigation of the crystal environment of ammonia. Acta Cryst A58:244–251

Bytheway I, Grimwood DJ, Jayatilaka D (2002) Wavefunctions derived from experiment. III. Topological analysis of crystal fragments. Acta Cryst A58:232–243

Capelli SC, Bürgi HB, Dittrich B, Grabowsky S, Jayatilaka D (2014) Hirshfeld Atom Refinement. Iucrj 1:361–379

Casati N, Genoni A, Meyer B, Krawczuk A, Macchi P (2017) Exploring charge density analysis in crystals at high pressure: data collection, data analysis and advanced modelling. Acta Cryst Sect B 73:584–597

Cassam-Chenaï P, Jayatilaka D (2001) Some fundamental problems with zero flux partitioning of electron densities. Theor Chem Acc 105:213–218

Christenson JH, Cronin JW, Fitch VL, Turlay R (1964) Evidence for the 2π decay of the K_2^0 meson system. Phys Rev Lett 13:138–140

Clementi E, Raimondi DL (1963) Atomic Screening constants from SCF functions. J Chem Phys 38:2686–2689

Clementi E, Roetti C (1974) Roothaan-Hartree-Fock atomic wavefunctions: Basis functions and their coefficients for ground and certain excited states of neutral and ionized atoms, $Z \leq 54$. At Data Nucl Data Tables 14:177–478

Clinton WL, Nakhleh J, Wunderlich F (1969) Direct determination of pure-state density matrices. I. Some simple introductory calculations. Phys Rev 177:1–6

Clinton WL, Galli AJ, Massa LJ (1969) Direct determination of pure-state density matrices. II. Construction of constrained idempotent one-body densities. Phys Rev 177:7–13

Clinton WL, Henderson GA, Prestia JV (1969) Direct determination of pure-state density matrices. III. Purely theoretical densities via an electrostatic-virial theorem. Phys Rev 177:13–18

Clinton WL, Lamers GB (1969) Direct determination of pure-state density matrices. IV. Investigation of another constraint and another application of the P equations. Phys Rev 177:19–27

Clinton WL, Galli AJ, Henderson GA, Lamers GB, Massa LJ, Zarur J (1969) Direct determination of pure-state density matrices. V. Constrained eigenvalue problems. Phys Rev 177:27–33

Clinton WL, Massa LJ (1972) Determination of the electron density matrix from X-ray diffraction data. Phys Rev Lett 29:1363–1366

Clinton WL, Frishberg CA, Massa LJ, Oldfield PA (1973) Methods for obtaining an electron-density matrix from X-ray diffraction data. Int J Quantum Chem 7:505–514

Compton AH (1915) The distribution of the electrons in atoms. Nature 95:343–344

Compton AH (1935) X-rays in theory and experiment. Macmillan and Co., London

Compton AH (1965) X-rays as a branch of optics. In: Nobel lectures, physics 1922–1941. Elsevier Publishing Company, Amsterdam.

Cooper MJ (1985) Compton scattering and electron momentum determination. Rep Prog Phys 48:415–481

Cooper MJ (2016) The Saga of Sagamore. Physica Scripta 91:012501

Cooper MJ, Mijnarends PE, Shiotani N, Sakai N, Bansil A (2004) X-ray compton scattering from electrons. Oxford University Press, Oxford

Coppens P (1967) Comparative X-ray and neutron diffraction study of bonding effects in s-triazine. Science 158:1577–1579

Coppens P (1984) Can we see the electrons? J Chem Educ 61:761–765

Coppens P (1997) X-ray charge densities and chemical bonding. Oxford University Press, Oxford

Coppens P, Csonka LN, Willoughby TV (1970) Electron population parameters from least-squares refinement of X-ray diffraction data. Science 167:1126–1128

Coppens P, Pautler D, Griffin JF (1971) Electron population analysis of accurate diffraction data. II application of one-center formalisms to some organic and inorganic molecules. J Am Chem Soc 93:1051–1058

Coppens P, Willoughby TV, Csonka LN (1971) Electron population analysis of accurate diffraction data. I. Formalisms and restrictions. Acta Cryst A27:248–256

Coulthard MA (1967) A relativistic Hartree-Fock atomic field calculation. Proc Phys Soc 91:44–49

Craven BM, Weber HP, He X (1987) The POP Least Squares Refinement Procedure. Technical Report TR-872; Department of Crystallography, University of Pittsburgh: Pittsburgh, PA.

Cremer D, Kraka E (1984) A description of the chemical-bond in terms of local properties of electrondensity and energy. Croat Chem Acta 57:1259–1281

Danovich D, Shaik S, Rzepa SH, Hoffmann R (2013) A response to the critical comments on "one molecule, two atoms, three views, four bonds?" Angew Chem Int Ed 52:5926–5928

Dawson B (1964) Aspherical atomic scattering factors in crystal structure refinement i. coordinate and thermal motion effects in a model centrosymmetric system. Acta Cryst 17:990–996

Dawson B (1967) A general structure factor formalism for interpreting accurate X-ray and neutron diffraction data. Proc Royal Soc A 298:255–263

Debye P (1913) Interferenz von Röntgenstrahlen und Wärmebewegung. Ann Phys 348:49–92

Debye P (1915) Zerstreuung Von Röntgenstrahlen. Ann Phys 351:809–823

Debye P (1930) Röntgeninterferenzen Und Atomgrösse. Phys Z 31:419–428

Destro R, Roversi P, Barzaghi M, Marsch RE (2000) Experimental charge density of α-glycine at 23 K. J Phys Chem A 104:1047–1054

Deutsch M, Claiser N, Pillet S, Chumakov Y, Becker P, Gillet JM, Gillon B, Lecomte C, Souhassou M (2012) Experimental determination of spin-dependent electron density by joint refinement of X-ray and polarized neutron diffraction data. Acta Cryst A68:675–686

Deutsch M, Gillon B, Claiser N, Gillet JM, Lecomte C, Souhassou M (2014) First spin-resolved electron distributions in crystals from combined polarized neutron and X-ray diffraction experiments. IUCrJ 1:194–199

Dirac PAM (1929) The quantum mechanics of many-electron systems. Proc R Soc Lond A123:714–733

Dittrich B (2017) Is there a future for topological analysis in experimental charge-density research? Acta Cryst B73:325–329

Dittrich B, Koritsanszky T, Luger P (2004) A simple approach to nonspherical electron densities by using invarioms. Angew Chem Int Ed Engl 43:2718–2721

Domagała S, Fournier B, Liebschner D, Guillot B, Jelsch C (2012) An improved experimental data-bank of transferable multipolar atom models–ELMAM2. Construction details and applications. Acta Cryst A63:108–125

Dos Santos LHR, Genoni A, Macchi P (2014) Unconstrained and X-ray constrained extremely localized molecular orbitals: analysis of the reconstructed electron density. Acta Cryst A70:532–551

Dunitz JD, Schweizer WB, Seiler P (1983) X-ray study of the deformation density in tetrafluo-roterephthalodinitrile: weak bonding density in the C, F-bond. Helv Chim Acta 66:123–133

Einstein A (1905) Zur Elektodynamik Bewegter Körper. Ann Phys 17:891–921

Einstein A (1916) Grundlage Der Allgemeinen Relativitätstheorie. Ann Phys 49:769–822

Eisenschitz R, London F (1930) Über das Verhältnis der van der Waalsschen Kräfte zu den homöopolaren Bindungskräften. Z Physik 60:491–527

Ernst M, Dos Santos LHR, Krawczuk A, Macchi P (2019) Towards a generalized database of atomic polarizabilities. In: Chopra D (ed) Understanding intermolecular interactions in crystals, vol 2019, pp 211–242.

Ernst M, Genoni A, Macchi P (2020) Analysis of crystal field effects and interactions using X-ray restrained ELMOs. J Mol Struct 1209:127975

Espinosa E, Molins E, Lecomte C (1998) Hydrogen bond strengths revealed by topological analyses of experimentally observed electron densities. Chem Phys Lett 285:170–173

Farrugia LJ, Evans C (2005) Experimental X-ray charge density studies on the binary carbonyls $Cr(CO)_6$, $Fe(CO)_5$, and $Ni(CO)_4$. J Phys Chem 109:8834–8848

Feynman RP (1939) Forces in molecules. Phys Rev 56:340–343

Figgis BN (1966) Introduction to ligand fields. Interscience, New York

Figgis BN, Reynolds PA, Williams GA (1980) Spin density and bonding in the $CoCl_4^{2-}$ ion in Cs_3CoCl_5. Part 2. Valence electron distribution in the $CoCl_4^{2-}$ ion. J Chem Soc, Dalton Trans 2339–2347

Fischer A, Tiana D, Scherer W, Batke K, Eickerling G, Svendsen H, Bindzus N, Iversen BB (2011) Experimental and theoretical charge density studies at subatomic resolution. J Phys Chem A 115:13061–13071

Fock VA (1930) Näherungsmethode zur Lösung des quantenmechanischen Mehrkörperproblems. Z Phys 61:126–148

Friedrich W, Knipping P, Laue M (1912) Interferenz-Erscheinungen bei Röntgenstrahlen. Sitzungs-berichte der Kgl. Bayer. Akad. der Wiss 303–322

Fugel M, Jayatilaka D, Hupf E, Overgaard J, Hathwar VH, Macchi P, Turner MJ, Howard JAK, Dolomanov O, Puschmann H, Iversen BB, Bürgi HB, Grabowsky S (2018) Probing accuracy and precision of Hirshfeld Atom Refinement with HARt interfaced to Olex2. IUCrJ 5:32–44

Grabowsky S, Luger P, Buschmann J, Schneider Th, Schirmeister T, Sobolev AN, Jayatilaka D (2012) The significance of ionic bonding in sulfur dioxide: bond orders from X-ray diffraction data. Angew Chem 51:6776–6779

Gatti C, Macchi P (eds) (2012) Modern charge density analysis. Springer, Dordrecht

Gatti C, Saunders VR, Roetti C (1994) Crystal field effects on the topological properties of the electron density in molecular crystals: The case of urea. J Chem Phys 101:10686–10696

Gatti C (2005) Chemical bonding in crystals: new directions. Z Kristall 220:399–457

Gautier R, Gautier R, Hernandez O, Audebrand N, Bataille T, Roiland C, Alkaim E, Le Pollès L, Furet E, Le Fur E (2013) DFT-assisted structure determination of α_1- and α_2-VOPO$_4$: new insights into the understanding of the catalytic performances of vanadium phosphates. Dalton Trans 42:8124–8131

Genoni A (2013a) Molecular orbitals strictly localized on small molecular fragments from X-ray diffraction data. J Phys Chem Lett 4:1093–1099

Genoni A (2013b) X-ray constrained extremely localized molecular orbitals: theory and critical assessment of the new technique. J Chem Theory Comput 9:3004–3019

Genoni A (2017) A first-prototype multi-determinant X-ray constrained wavefunction approach: the X-ray constrained extremely localized molecular orbital-valence bond method. Acta Cryst 73:312–316

Genoni A, Bucinsky L, Claiser N, Contreras-Garcia J, Dittrich B, Dominiak PM, Espinosa E, Gatti C, Giannozzi P, Gillet JM, Jayatilaka D, Macchi P, Madsen AØ, Massa LJ, Matta CF, Merz KM, Nakashima PNH, Ott H, Ryde U, Schwarz K, Sierka M, Grabowsky S (2018) Quantum crystallography: current developments and future perspectives. Chem Eur J 24:10881–10905

Genoni A, Macchi P (2020) Quantum crystallography in the last decade: developments and outlooks. Crystals 10:476

Gerlach W, Stern O (1922) Der experimentelle Nachweis der Richtungs-quantelung im Magnetfeld. Z Physik 9:349–352

Ghermani N-E, Bouhmaida N, Lecomte C (1993) Modelling electrostatic potential from experimentally determined charge densities. I. Spherical-atom approximation. Acta Cryst A49:781–789

Giacovazzo C, Monaco HL, Artioli G, Viterbo D, Milanesio M, Ferraris G, Gilli G, Gilli P, Zanotti G, Catti M (2016) Fundamentals of crystallography. IUCr texts on crystallography. Oxford University Press, Oxford

Gilbert TL (1975) Hohenberg-Kohn theorem for nonlocal external potentials. Phys Rev B 12:2111–2120

Gillet JM (2007) Determination of a one-electron reduced density matrix using a coupled pseudo-atom model and a set of complementary scattering data. Acta Cryst A63:234–238

Gillet JM, Becker P, Cortona P (2001) Joint refinement of a local wave-function model from Compton and Bragg scattering data. Phys Rev B 63:235115

Gordon W (1926) Der Comptoneffekt nach der Schrödingerschen Theorie. Z Phys 40:117–133

Grabowsky S, Genoni A, Bürgi HB (2017) Quantum crystallography. Chem Sci 8:4159–4176

Grabowsky S, Genoni A, Thomas SP, Jayatilaka D (2020) The advent of quantum crystallography: form and structure factors from quantum mechanics for advanced structure refinement and wavefunction fitting. Struct Bond 186:6–144

Grell J, Bernstein J, Tinhofer G (2002) Investigation of hydrogen bond patterns: a review of mathematical tools for the graph set approach. Cryst Rev 8:1–56

Griffiths D (1987) Introduction to elementary particles. Wiley, New York

Griffiths D (2016) Introduction to quantum mechanics. Cambridge University Press, Cambridge

Grimwood D, Bytheway I, Jayatilaka D (2003) Wave functions derived from experiment. V. Investigation of electron densities, electrostatic potentials, and electron localization functions for noncentrosymmetric crystals. J Comput Chem 24:470–483

Grimwood D, Jayatilaka D (2001) Wavefunctions derived from experiment. I. A wavefunction for oxalic acid dihydrate. Acta Cryst A57:87–100

Gruza B, Chodkiewicz ML, Krzeszczakowska J, Dominiak PM (2020) Refinement of organic crystal structures with multipolar electron scattering factors. Acta Cryst A76:92–109

Guinier A (1963) X-ray diffraction in crystals, imperfect crystals, and amorphous bodies. Dover Publications Inc., New York

Guillot B, Jelsch C, Podjarny A, Lecomte C (2008) Charge-density analysis of a protein structure at subatomic resolution: the human aldose reductase case. Acta Cryst D64:567–588

Hansen NK, Coppens P (1978) Electron population analysis of accurate diffraction data. 6. Testing aspherical atom refinements on small-molecule data sets. Acta Cryst A34:909–921

Hartree RD (1928a) The wave mechanics of atom with a non-Coulomb central field. Part I Theory and methods. Math Proc Cambr Phil Soc 24:89–110

Hartree RD (1928b) The wave mechanics of an atom with a non-coulomb central field. Part II Some results and discussion. Math Proc Camb Philos Soc 24:111–132

Heisenberg W (1927) Über den anschaulichen Inhalt der quantentheoretischen Kinematik und Mechanik. Z Physik 43:172–198

Hellmann H (1937) Einführung in die Quantenchemie. F. Deuticke, Leipzig

Hibbs DE, Howard ST, Huke JP, Waller MP (2005) A new orbital-based model for the analysis of experimental molecular charge densities: an application to (Z)-N-methyl-C-phenylnitrone. Phys Chem Chem Phys 7:1772–1778

Hirshfeld FL, Rzotkiewicz S (1974) Electrostatic binding in the first-row AH and A_2 diatomic molecules. Mol Phys 27:1319–1343

Hirshfeld FL (1971) Difference densities by least-squares refinement: fumaramic acid. Acta Cryst B27:769–781

Hirshfeld FL (1976) Can X-ray data distinguish bonding effects from vibrational smearing? Acta Cryst A32:239–244

Hirshfeld FL (1977a) Charge deformation and vibrational smearing. Isr J Chem 16:168–174

Hirshfeld FL (1977b) Bonded-atom fragments for describing molecular charge-densities. Theor Chim Acta 44:129–138

Hirshfeld FL (1984) Hellmann-feynman constraint on charge densities, an experimental test. Acta Cryst B40:613–615

Hohenberg P, Kohn W (1964) Inhomogeneous electron gas. Phys Rev B 136:864–871

Hofmann A, Netzel J, van Smaalen S (2007) Accurate charge density of trialanine: a comparison of the multipole formalism and the maximum entropy method (MEM). Acta Cryst B63:285–295

Holladay A, Leung PC, Coppens P (1983) Generalized relations between d-orbital occupancies of transition-metal atoms and electron-density multipole population parameters from X-ray diffraction data. Acta Cryst A39:377–387

Hopf H (1926) Vektorfelder in n-Dimensionalen Mannigfaltigkeiten. Math Ann 96:209–221

Hübschle CB, Van Smaalen S (2017) The electrostatic potential of dynamic charge densities. J App Cryst 50:1627–1636

Hupf E, Olaru M, Raţ CI, Fugel M, Hübschle CB, Lork E, Grabowsky S, Mebs S, Beckmann J (2017) Mapping the trajectory of nucleophilic substitution at silicon using a peri-substituted acenaphthyl scaffold. Chem Eur J 44:10568–10579

Iversen BB, Larsen FK, Souhassou M, Takata M (1995) Experimental evidence for the existence of non-nuclear maxima in the electron-density distribution of metallic beryllium. A comparative study of the maximum entropy method and the multipole refinement method. Acta Cryst B51:580–591

James RW (1958) The optical principles of the diffraction of X-rays. G. Bell and Sons, London

James RW, Brindley GW (1931) Some numerical calculations of atomic scattering factors. Phyl Mag 12:81–112

Jarzembska KN, Dominiak PM (2012) New version of the theoretical databank of transferable aspherical pseudoatoms, UBDB2011-towards nucleic acid modelling. Acta Cryst A68:139–147

Jayatilaka D (1998) Wave function for beryllium from X-ray diffraction data. Phys Rev Lett 80:798–801

Jayatilaka D, Dittrich B (2008) X-ray structure refinement using aspherical atomic density functions obtained from quantum-mechanical calculations. Acta Cryst A64:383–393

Jayatilaka D, Grimwood D (2001) Wavefunctions derived from experiment. I. Motivation and theory. Acta Cryst A57:76–86

Jaynes ET (1968) Prior probabilities. IEEE Trans Syst Sci Cybern 4:227–240

Jelsch C, Teeter MT, Lamzin V, Pichon-Pesme V, Blessing RH, Lecomte C (2000) Accurate protein crystallography at ultra-high resolution: Valence electron distribution in crambin. PNAS 97:3171–3176

Jelsch C, Guillot B, Lagoutte A, Lecomte C (2005) Advances in protein and small-molecule charge-density refinement methods using MoPro. J Appl Cryst 38:38–54

Jeffrey GA, Piniella J (1991) Application of charge density research to chemistry and drug design. NATO ASI Series, Series B: Physics, vol 250

Jones W, March NH (1985) Theoretical solid state physics, vol 1 Appendix A1.6, Dover, New York.

Jost A, Rees B, Yelon WB (1975) Electronic structure of chromium hexacarbonyl at 78 K. I. Neutron diffraction study. Acta Cryst B31:2649–2658

Kibalin I, Voufack AB, Souhassou M, Gillon B, Gillet J-M, Claiser N, Gukasov A, Porcher F, Lecomte C (2021) Spin-resolved atomic orbital model refinement for combined charge and spin density analysis: application to the $YTiO_3$ perovskite. Acta Cryst A77:96–104

Klein O (1926) Quantentheorie Und Fünfdimensionale Relativitätstheorie. Z Phys 37:895–906

Koritsanszky TS, Coppens P (2001) Chemical applications of X-ray charge-density analysis. Chem Rev 101:1583–1627

Koritsanszky T, Howard ST, Richter T, Mallinson PR, Su Z, Hansen NK (1995) XD. A computer program package for multipole refinement and analysis of charge densities from X-ray diffraction data. Free University of Berlin.

Krawczuk A, Macchi P (2014) Charge density analysis for crystal engineering. Chem Cent J 8:68

Kumar P, Gruza B, Bojarowski SA, Dominiak PM (2018) Extension of the transferable aspherical pseudoatom data bank for the comparison of molecular electrostatic potentials in structure-activity studies. Acta Cryst A7:398–408

Kunze KL, Hall MB (1986) Why the accumulation of electron density appears weak or absent in certain covalent bonds. J Am Chem Soc 108:5122–5127

Kurki-Suonio K, Meisalo V (1966) Nonspherical deformations of the ions in Fluorite. J Phys Soc Jpn 21:122–126

Kurki-Suonio K (1968) On the information about deformations of the atoms in X-ray diffraction data. Acta Cryst A24:379–390

Leung PC, Coppens P (1983) Experimental charge density study of Dicobalt Octacarbonyl and comparison with theory. Acta Cryst B39:535–542

Lewis GN (1916) The Atom and the Molecule. J Am Chem Soc 38:762–785

Lübben J, Wandtke CM, Hübschle CB, Ruf M, Sheldrick GM, Dittrich B (2019) Aspherical scattering factors for SHELXL–model, implementation and application. Acta Cryst A75:50–62

Lundeen JS, Sand B, Patel A, Stewart C, Bamber C (2011) Direct measurement of the quantum wavefunction. Nature 474:188–191

Macchi P (2017) The future of topological analysis in experimental charge-density research. Acta Cryst B73:330–336

Macchi P (2020) The connubium between crystallography and quantum mechanics. Cryst Rev 26:209–268

Macchi P (2022) Quantum crystallography: fundamentals and applications. De Gruyter, to be published

Macchi P, Coppens P (2001) Relativistic analytical wave functions and scattering factors for neutral atoms beyond Kr and for all chemically important ions up to I-. Acta Cryst A57:656–662

Macchi P, Garlaschelli L, Sironi A (2002) Electron density of semi-bridging carbonyls. Metamorphosis of CO ligands observed via experimental and theoretical investigations on $[FeCo(CO)_8]^-$. J Am Chem Soc 124:14173–14184

Macchi P, Sironi A (2003) Chemical bonding in transition metal carbonyl clusters: complementary analysis of theoretical and experimental electron densities. Coord Chem Rev 238–239:383–412

Malaspina LA, Wieduwilt EK, Bergmann J, Kleemiss F, Meyer B, Ruiz-López MF, Pal R, Hupf E, Beckmann J, Piltz RO, Edwards AJ, Grabowsky S, Genoni A (2019) Fast and accurate quantum crystallography: from small to large, from light to heavy. J Phys Chem Lett 10:6973–6982

Martin M, Rees B, Mitschler A (1982) Bonding in a binuclear metal carbonyl: experimental charge density in $Mn_2(CO)_{10}$. Acta Cryst B38:6–15

Massa LJ, Matta CF (2021) Quantum crystallography. De Gruyter, Berlin

Matthewman JC, Thompson P, Brown PJ (1982) The CCSL library. J Appl Cryst 15:167–173

McLean AD, McLean RS (1981) Roothaan-Hartree-Fock atomic wave functions Slater basis-set expansions for Z = 55–92. At Data Nucl Data Tables 26:197–381

McWeeny R (1951) X-ray scattering by aggregates of bonded atoms. I. Analytical approximations in singe-atom scattering. Acta Cryst 4:513–519

McWeeny R (1952) X-ray scattering by aggregates of bonded atoms. II. The effect of the bonds: with an application to H_2. Acta Cryst 5:463–468

McWeeny R (1953) X-ray scattering by aggregates of bonded atoms. III. The bond scattering factor: simple methods of approximation in the general case. Acta Cryst 6:631–637

McWeeny R (1954) X-ray scattering by aggregates of bonded atoms. IV. Applications to the carbon atom. Acta Cryst 7:180–185

Miao M-S, Hoffmann R (2014) High pressure electrides: a predictive chemical and physical theory. Acc Chem Res 47:1311–1317

Michel K, Meredig B, Ward L, Wolverton C (2019) First-principles-assisted structure solution: leveraging density functional theory to solve experimentally observed crystal structures. In: Andreoni W, Yip S (eds) Handbook of materials modeling. Springer, Cham

Minkin VI (1991) Glossary of terms used in theoretical organic chemistry. Pure Appl Chem 71:1919–1981

Morse M (1925) Relations between the critical points of a real function of n independent variables. Trans Am Math Soc 27:345–396

Mukherji A, Karplus M (1963) Constrained molecular wavefunctions: HF molecule. J Chem Phys 38:44–48

Netzel J, van Smaalen S (2009) Topological properties of hydrogen bonds and covalent bonds from charge densities obtained by the maximum entropy method (MEM). Acta Cryst B65:624–638

Nielsen MA, Chuang IL (2010) Quantum computation and quantum information. Cambridge University Press, Cambridge

Palatinus L, van Smaalen S (2005) The prior-derived F constraints in the maximum-entropy method. Acta Cryst A61:363–372

Palke WE (1986) Double bonds are bent equivalent hybrid (banana) bonds. J Am Chem Soc 108:6543–6544

Pauling L (1926) Letter to A.A. Noyes. Linus Pauling and the nature of the chemical bond: a documentary history website. http://scarc.library.oregonstate.edu/coll/pauling/bond/corr/corr278.1-lp-noyes-19261217.html

Pauling L (1939) The nature of the chemical bond. Cornell University Press, Ithaca

Perdew JP, Burke K, Ernzerhof M (1996) Generalized gradient approximation made simple. Phys Rev Lett 77:3865

Petricek V, Dusek M, Palatinus L (2014) Crystallographic computing system JANA2006: general features. Z Kristallogr 229:345–352

Pichon-Pesme V, Lecomte C, Lachekar H (1995) On building a data bank of transferable experimental electron density parameters applicable to polypeptides. J Phys Chem 99:6242–6250

Popelier PLA, Aicken FM (2003) Atomic properties of amino acids: computed atom types as a guide for future force field design. ChemPhysChem 4:824–829

Popelier PLA (2016) Quantum chemical topology. In: Mingos M (ed) The Chemical Bond–100 years old and getting stronger. Structure and Bonding, vol 170, pp 71–117

Popper KR (1982) Quantum theory and the schysm in physics. Unwin Hyman, London

Price SL (2018) Control and prediction of the organic solid state: a challenge to theory and experiment. Proc Royal Soc A 474:20180351

Pyykko P, Desclaux JP (1979) Relativity and the periodic system of elements. Acc Chem Res 12:276–281

Quack M, Stohner J, Willeke M (2008) High-resolution spectroscopic studies and theory of parity violation in chiral molecules. Annu Rev Phys Chem 59:741–769

Rietveld HM (1969) A profile refinement method for nuclear and magnetic structures. J App Cryst 2:65–71

Rose ME (1957) Elementary Theory of Angular Momentum. Wiley, New York

Roversi P, Merati F, Barzaghi M, Destro R (1996) Charge density in crystalline citrinin from X-ray diffraction at 19 K. Can J Chem 74(1145):1161

Roversi P, Irwin JJ, Bricogne G (1998) Accurate charge-density studies as an extension of Bayesian crystal structure determination. Acta Cryst A54:971–996

Salam A (1991) The role of chirality in the origin of life. J Mol Evol 33:105–113

Salam A (1992) Chirality, phase transitions and their induction in amino acids. Phys Lett B 288:153–160

Savin A, Flad HJ, Flad J, Preuss H, von Schnering HG (1992) On the bonding in carbosilanes. Angew Chem Int Ed Engl 31:185–187

Scatena R, Guntern YT, Macchi P (2019) Electron density and dielectric properties of highly porous MOFs: binding and mobility of guest molecules in $Cu_3(BTC)_2$ and $Zn_3(BTC)_2$. J Am Chem Soc 141:9382–9390

Schmider H, Smith VH, Werich W (1993) On the inference of the one-particle density matrix from position and momentum-space form factors. Z Naturforsh 48a:211–220

Schwarzenbach D (1966) Elektrische feldgradienten und die kovalenz der bindungen in $AlPO_4$. Z Natur 123:422–442

Schwinger J (1951a) On the Green's functions of quantized fields I. In: Proceedings of the National Academy of Science of the United States of America, vol 37, pp 452–455

Schwinger J (1951b) On the Green's functions of quantized fields II. In: Proceedings of the National Academy of Science of the United States of America, vol 37, pp 455–459

Shevchenko AP, Blatov VA (2021) Simplify to understand: how to elucidate crystal structures? Struct Chem 32:507–519

Schomaker V, Trueblood KN (1968) On the rigid-body motion of molecules in crystals. Acta Cryst B24:63–76

Spackman MA, Maslen EN (1985) Electron density and the chemical bond. a reappraisal of Berlin's Theorem. Acta Cryst A41:347–353

Stash A, Tsirelson VG (2002) WinXPRO: a program for calculating crystal and molecular properties using multipole parameters of the electron density. J App Cryst 35:371–373

Stash A, Tsirelson VG (2014) Developing WinXPRO: a software for determination of the multipole-model-based properties of crystals. J App Cryst 47:2086–2089

Stevens ED (1980) Relationship of multipole populations to d-orbital occupancies of a transition metal atom in a tetragonally distorted octahedral field. In: Becker, PJ (ed) Electron and magnetization densities in molecules and crystals, pp 823–826

Stevens ED, Coppens P (1979) Refinement of metal d-orbital occupancies from X-ray diffraction data. Acta Cryst A35:536–539

Stewart RF (1969) Generalized X-ray scattering factors. J Phys Chem 51:4569–4577

Stewart RF (1973) Electron population analysis with generalized X-ray-scattering factors–higher multipoles. J Chem Phys 58:1668–1676

Stewart RF, Bentley J, Goodman B (1975) Generalized X-ray scattering factors in diatomic molecules. J Chem Phys 63:3786–3793

Stewart RF (1976) Electron population analysis with rigid pseudoatoms. Acta Cryst A32:565–574

Stewart RF (1979) On the mapping of electrostatic properties from bragg diffraction data. Chem Phys Lett 65:335–342

Stewart RF (1982) Mapping electrostatic potentials from diffraction data. God Jugosl Cent Kristalogr 17:1–24

Stewart RF, Spackman MA (1983) Valray user's manual. Department of Chemistry, Carnegie-Mellon University, Pittsburgh, PA

Stewart RF, Spackman MA, Flensburg C (2000) VALRAY user's manual. Carnegie Mellon University and University of Copenhagen

Streltsov VA, Nakashima PNH, Johnson AWS (2001) Charge density analysis from complementary high energy synchrotron X-ray and electron diffraction data. J Phys Chem Solid 62:2109–2117

Su Z, Coppens P (1992) On the mapping of electrostatic properties from the multipole description of the charge density. Acta Cryst A48:188–197

Su Z, Coppens P (1998) nonlinear least-squares fitting of numerical relativistic atomic wave functions by a linear combination of slater-type functions for atoms with $Z = 1$–36. Acta Cryst A54:646–652

Tanaka K (2018) X-ray molecular orbital analysis. I. Quantum mechanical and crystallographic framework. Acta Cryst A74:345–356

Tanaka K, Makita R, Funahashi S, Komori T, Win Z (2008) X-ray atomic orbital analysis. I. Quantum mechanical and crystallographic framework of the method. Acta Cryst B64:437–449

Tanaka K, Wasada Tsutsui Y (2021) X-ray molecular orbital analysis. II. Application to diformohydrazide, $(NHCHO)_2$. Acta Cryst A77:593–610

Takata M, Sakata M (1996) The influence of the completeness of the data set on the charge density obtained with the maximum-entropy method. A re-examination of the electron-density distribution in Si. Acta Cryst A52:287–290

Thomas LH, Umeda K (1957) Atomic scattering factors calculated from the TFD atomic model. J Chem Phys 26:293

Tsirelson VG, Ozerov RP, Zavodnik VE, Fomicheva EB, Kuznetsova LI, Rez IS (1989) Valence electron distribution in crystals. II. Calculation of the electronstructure of lithium formate deuterate from the diffraction data. Sov Phys Crystallogr 25:735–742

Tsirelson VG, Ozerof RP (1995) Electron density and bonding in crystals. Institute of Physics Publishing, Bristol and Philadelphia

Tsirelson VG (2002) The mapping of electronic energy distributions using experimental electron density. Acta Cryst B58:632–639

Uhlenbeck GE, Goudsmit S (1925) Ersetzung der Hypothese vom unmechanischen Zwang durch eine Forderung bezüglich des inneren Verhaltens jedes einzelnen Elektrons. Naturwissenschaften 13:953–954

Van Smaalen S, Netzel J (2009) The maximum entropy method in accurate charge-density studies. Physica Scripta 79:048304

Van Vleck J (1932) Theory of the variations in paramagnetic anisotropy among different salts of the iron group. Phys Rev 41:208–215

Verschoor GC (1964) X-ray diffraction study of the electron density distribution in cyanuric acid. Nature 202:1206–1207

Vogel K, Risken H (1989) Determination of quasiprobability distributions in terms of probability distributions for the rotated quadrature phase. Phys Rev A 40: 2847–2849.

Volkov A, Li X, Koritsanszky T, Coppens P (2004) Ab initio quality electrostatic atomic and molecular properties including intermolecular energies from a transferable theoretical pseudoatom databank. J Phys Chem A 108:4283–4300

Volkov A, Macchi P, Farrugia LJ, Gatti C, Mallinson P, Richter T, Koritsanszky TS (2006) XD2006-a computer program package for multipole refinement, topological analysis of charge densities and evaluation of intermolecular energies from experimental and theoretical structure factors. The State University of New York at Buffalo

Volkov A, Macchi P, Farrugia LJ, Gatti C, Mallinson P, Richter T, Koritsanszky TS (2016) XD2016-a computer program package for multipole refinement, topological analysis of charge densities and evaluation of intermolecular energies from experimental and theoretical structure factors

Wall ME (2016) Quantum crystallographic charge density of urea. IUCrJ 3:237–246

Waller I (1923) Zur Frage der Einwirkung der Wärmebewegung auf die Interferenz von Röntgenstrahlen. Z Phys A 17:398–408

Wang JH, Sagar RP, Schmider H, Smith VH (1993) X-ray elastic and inelastic scattering factors for neutral atoms Z = 2–92. Atomic Data Nucl Tables 53:233–269

Weiss RJ, Demarco JJ (1958) X-ray determination of the number of 3d electrons in Cu, Ni Co, Fe, and Cr. Rev Mod Phys 30:59–62

Weiss RJ (1966) X-ray determination of electron distributions. North-Holland Publishing Company, Amsterdam

Welker J, Giessibl FJ (2012) Revealing the angular symmetry of chemical bonds by atomic force microscopy. Science 336:444–449

Weinberg S (2004) The making of the standard model. Eur Phys J C 34:5–13

Wells AF (1977) Three-dimensional nets and polyhedra. Wiley-Interscience, New York

Weyrich W (1996) In: Pisani C (ed) Quantum mechanical Ab initio calculation of the properties of crystalline materials. Springer, Berlin, pp 245–272

Zuo JM, Kim M, O'Keeffe M, Spence JCH (1999) Direct observation of d-orbital holes and Cu-Cu bonding in Cu_2O. Nature 401:49–52

Printed in the United States
by Baker & Taylor Publisher Services